I0468450

EFFECT OF SNAKE *NAJA NAJA* VENOM ON HAEMATOLOGICAL, SELECTED BIOCHEMICAL PARAMETERS AND HISTOPATHOLOGICAL STUDIES IN WISTAR STRAIN RATS

By

DR.M.MALLESWARI

U.G.C Post Doctoral Fellow

Dept. of Zoology

S.V.U. College of Sciences

SRI VENKATESWARA UNIVERSITY

TIRUPATI-517502

Dr. P.JOSTHNA

Assistant Professor

Dept.of Biotechnology

SRI PADMAVATHI MAHILA VISVAVIDYALAYAM

(Women's University)

TIRUPATI-517502

Prof.P.JACOB DOSS

Head, Dept. of Zoology

S.V.U. College of Sciences

SRI VENKATESWARA UNIVERSITY

TIRUPATI-517502

This book presents an overview of the **EFFECT OF SNAKE *NAJA NAJA* VENOM ON HAEMATOLOGICAL, SELECTED BIOCHEMICAL PARAMETERS AND HISTOPATHOLOGICAL STUDIES IN WISTAR STRAIN RATS.** Snake bite injuries and deaths are socio-medical problems of considerable magnitude. In India a large number of people suffer and die every year due to snake venom poisoning. Snake venom, though greatly feared, is a natural biological resource, containing several components that could be of potential therapeutic value. Use of snake venom in different patho-physiological conditions has been mentioned in Ayurveda, Homeopathy and Folk medicine. It is well known that snake venom is complex mixture of enzymes, peptides and proteins of low molecular mass with specific chemical and biological activities. Snake venom contains several neurotoxic, cardiotoxic, cytotoxic, nerve growth factor, lectins, disintrigrins, haemorrhagins and many other different enzymes. These proteins not only inflict death to animals and humans, but can also be used for the treatment of thrombosis, arthritis, cancer and many other diseases. An overview of various components of snake venom have prospects in health and disease management.

So far the mechanisms a few components are understood. Further studies in this area are essential, since the mechanisms of these proteins will contribute to our understanding of not only the toxicity of the venom components, but also the normal physiological events of haemostasis. Studies on the structure and functional relationships are very useful. Such studies will undoubtedly help in the development of different drugs, which might be useful for several medical applications.

This work is a modest attempt by the author towards an understanding the toxic potentials of snake *Naja naja* venom in different tissues of *albino rats*. This study is far from being comprehensive, yet the author remains hopeful that the present study would contribute useful information to the existing knowledge on changes in metabolic activities in protein metabolism during the effect of snake *Naja naja* venom. The author remains pardonable for any error which may have crept in due to oversight and for any investigative lacunae which are due to limitations in facilities and infrastructure.

Appropriate tools are employed for analysis and interpretation of the data. This thesis is composed of five chapters. The first chapter deals with the determination of lethal dose, LD_{50} by oral administration of snake *Naja naja* venom. The second chapter deals with alterations in hematological profiles. The third chapter deals with enzymes of protein metabolism and Identification of DNA damage. The fourth chapter deals with detoxification enzymes. The fifth chapter deals with the histopathological changes by light microscopy in different tissues of albino rat. Suggestions were made which are useful to all those people who are actively involved at different stages of designing, planning, organizing, implementing and evaluating the work.

ABOUT AUTHORS

Dr. M. Malleswari did her M.Sc Biotechnology from Mysore University and did her M.Phil in Biotechnology in Sri Venkateswara University ,Tirupati and completed her Ph.D at Sri Padmavathi Mahila Visvavidyalayam (Womens University) Tirupati. Presently doing UGC Post Doctoral Fellowship on Snake Venom Toxicology in the Department of Zoology in Sri Venkateswara University, Tirupati. She has published eight research articles in reputed Journals, and attended several National and International conferences /Seminars / Workshops.

Dr. P. Josthna, Assistant professor, Department of Biotechnology and Co-Coordinator for TePP Outreach cum Cluster Innovation Centre (TOCIC), SPMVV. Dr.P. Josthna completed her post doctoral fellowship at Yale University, Cunnecticut and University of Texas Medical Branch, Galveston, USA. Dr. P. Josthna has been continuing her research activity in the field of Cancer Biology. She is a member of various bodies like Ethological Society of India, Indian Science Congress and A.P Akademi of Science. Dr. P. Josthna has 15 years of teaching experience, 20 years of research experience and guided two Ph.D students and presently guiding 6 research scholars. She has published 25 research articles in reputed Nationals and International Journals. She has completed one UGC major research project and presently handling DST - SERB project. For her credit she has received awards like Young scientist, Talented Biotechnologist, Best NSS Programme Officer and Best paper presentation awards.

Prof. P. Jacob Doss, Head, Department of Zoology and the co-ordinator of DST-FIST and Deputy co-ordinator to the UGC SAP program and co-ordinator for DDE Zoology. He is also the editorial board member and treasurer to the ISCAP Journal. He is the member of Board of studies, and also member to several Universities. He has 23 years of teaching / research experience and successfully guided 25 ph.D and 10 M.Phil students. He has published 90 articles in reputed journals and authored 1 book. There are 4 major research projects to his credit.

ACKNOWLEDGEMENTS

I wish to express my profound sense of gratitude to my research supervisor Dr. P. Josthna, Assistant Professor of Biotechnology, Sri Padmavathi Mahila Visvavidyalayam (Women's University) Tirupati, for suggesting me this problem. I take this opportunity to express my heartfelt thanks to my research supervisor for all that she has done to me and I consider working with her is the greatest privilege I ever had.

I am very much indebted to Prof. P.Jacob Doss, Professor of Zoology, S.V.University, Tirupati, for his valuable suggestions, expert guidance, critical analysis and excellent programming of the problem, elegance of encouragement, affectionate care and for valuable comments and encouragement throughout the present investigation.

It is a great pleasure to express my heartfelt gratitude and profound respect to Prof. P.Sreenivasulu Reddy, Chairman, BOS, Dept. of Zoology, S.V.University, Tirupati, for his valuable suggestions and encouragement though out the research work.

I record my deep sense of gratitude to Prof. K. Vijaya Lakshmi, Head, Department of Biotechnology, Sri Padmavathi Mahila Visvavidyalayam (Women's University) Tirupati.

I am highly thankful to Dr. R. Usha, Co-ordinator, Department of Biotechnology for her constructive suggestions and continuous encouragement in all stages in a successful way.

My deepest feeling of love and gratitude goes to my parents Dr. G. Varalakshmi and M. Sreeramulu who laid foundation and remained as a source of inspiration for my career. They encouraged me to continue moving forward when I thought I would falter. I thank them for all they provided me - support, faith, confidence and patience. Diction is not enough to express my gratitude and regards, without whose ever lasting love and help in every walk of life, this work would not have been possible.

I wish to express my love and heartful gratitude to my husband *Mr.U.Murali* and my lovely kids *U.Briznath* and *U.Hithaishi, G.Sai Sruthi Keerthana* for their love, co-operation, never ending support, encouragement and patience till the completion of the work.

I wish to express my warmest feeling of love to my brother and sisters, **Dr.M.Chitty babu, Dr. M.C.Malleswari, Dr. M.Prasanna Kumari, Dr. M.Jagruthi** and to my brother-in-law **Dr. G.Subramanyam, Dr .Prasanna Jayakumar** for their affection and willingness to always help me will be remembered with sincere gratitude.

The present work could not have been completed without the encouragement, assistance and co-operation given me from time to time by my research colleagues *Ms. Geetharathan, Mrs. Lakshmi Deepika, Ms. Priyanka, Ms .Sreedevi, Mrs. Sumathi, Ms. Sujatha, Ms. Rajyalakshmi, Ms. Mohana, Mrs. Sreelatha, Mrs. Prasanna Lakshmi, Mrs. Uma Kranthi, Ms. Jeevana, Ms. Devi Srilakshmi Kala, Ms. Aswini Kumari, and Mrs. Jayasudha.*

I express my sincere thanks to Mr. Karimulla, Sri Venkateswara Veterinary University, Tirupati , for providing the necessary microphotograph facility.

I express my special thanks to animal keeper Mr. Anjaneyulu Raju for his help during the research work.

I wish to extend my thanks to all the teaching and non-teaching Staff members of Biotechnology, Sri Padmavathi Mahila Visvavidyalayam (Women's University), Tirupati, for their solicitous and ready help.

Lastly, I thank all my well wishers and whomsoever, who has helped me the least way possible.

M. Malleswari......✍

PREFACE

Snake bite injuries and deaths are socio-medical problems of considerable magnitude. In India a large number of people suffer and die every year due to snake venom poisoning. Snake venom, though greatly feared, is a natural biological resource, containing several components that could be of potential therapeutic value. Use of snake venom in different patho-physiological conditions has been mentioned in Ayurveda, Homeopathy and Folk medicine. It is well known that snake venom is complex mixture of enzymes, peptides and proteins of low molecular mass with specific chemical and biological activities. Snake venom contains several neurotoxic, cardiotoxic, cytotoxic, nerve growth factor, lectins, disintrigrins, haemorrhagins and many other different enzymes. These proteins not only inflict death to animals and humans, but can also be used for the treatment of thrombosis, arthritis, cancer and many other diseases. An overview of various components of snake venom have prospects in health and disease management.

Venomous and poisonous animals are a significant cause of global morbidity and mortality. In snakes, venom used for defense is an evolutionary adaptation to immobilize the prey. Venoms are the secretary substances of the venomous animals, which are synthesized and stored in specific areas of their body. Snake venoms are a unique physiological product of nature as they are mixtures of different substances, which are highly specific and have great affinity for different crucial and essential functional organization of cells and tissues. Efforts are already on for the use of these natural resources as powerful probes for elucidating complex biological processes of vital importance. Several isolated snake venom proteins with known mode of action have found practical applications as pharmaceutical agents, diagnostic agents or preparative tools in haemostaseology, neurobiology and complement research. Varieties of snake venom toxins are of interest in drug design, because the snake venom toxins provide three-dimensional templates for creating small molecule mimicking interesting pharmacological properties.

The use of snake venom as medicine was known to man for centuries. It is over sixty years since it was first realized that the physiologically active components of snake venoms might have therapeutic potential. In *Charak Samhita,* cobra venom has been said to be useful in *Dushyodara* and *Jalodara* (ascities). *Sushruta* and *Vagbhata* also mentioned similar use. *Saranghara Samita* mentioned the use of cobra venom in *'Sannipatik Jwara'*. In the Unani system of medicine cobra venom has been used as a tonic, aphrodisiac, hepatic stimulant and for revival in collapsed conditions. Venoms of *Viper, Crotalus, Cobra* and *Lacasis* are also routinely used in homeopathic medicine. Chinese physicians use snake venom products routinely to treat stroke and view them as effective and relatively safe. Natural protease inhibitors to haemorrhagins in snake venoms and their potential use in medicine have also been reported.

Applications of snake venom in different fields of medicine

Fibrinogenolytic and fibrinolytic activity of snake venom

The ability of some snake venom enzymes to clot fibrinogen has resulted in great therapeutic importance. These enzymes remove fibrinogen from the circulation without converting it to fibrin, or causing platelet aggregation. Ancord and batroxobin have been investigated in patients with stroke, deep vein thrombosis and cerebral infarction, myocardial infarction, peripheral atrial thrombosis, priapism and sickle cell crisis etc.,

Cardiotonic and antiarrythmic activity of snake venom

When snake venom ancord was treated with heart patients, the patients condition has been greatly improved.

Antineoplastic activity of snake venom

Snake venom has been used to develop newer drugs to combat various diseases including cancer. Calmette *et al.,* (1933) investigated the use of cobra venom in the treatment of cancer in mice. Match (1936), Macht (1938) showed that cobra venom, in extremely minute doses produced analgesic effects. This led to the possibility of therapeutic use of the cobra venom in arthritis and cancer.

Antiparalytic activity of snake venom

Venoms of several snakes are known to cause muscular paralysis. Subsequently several neurotoxic components that inhibit neuromuscular transmission by attacking different targets have been isolated. Neurotoxins from snake venoms have been utilised in different pharmacological and biochemical studies of nicotinic acetylcholine receptors (nAChRs) in the neuron and neuromuscular junction.

Anti arthritic activity of snake venom

It was found that cobra venom factor pre-treatment of rabbits prevented the induction of experimental immune complex arthritis (Henson, 1975). Snake venom has been used to elucidate the pathophysiology of several experimental model of arthritis.

Other possible therapeutic action of snake venom

The use of snake venom as an analgesic agent is well documented (Picolo *et al.*, 1998). Two venom based medications, cobroxin and nyloxin were marketed for the treatment of pain arthritis and other disorders. The Indian King Cobra *Ophiophagus hannah* venom induces immunomodulatory and haemopoietic stimulant activity, (Pallabi, D.E, 2000). Recently from the venom of *Vipera russeli*, a heat stable compound has been identified which was found to produce cardiorespiratory modulation in animal models with a possible application in cardiorespiratory related pathophysiological condition.

So far the mechanisms a few components are understood. Further studies in this area are essential, since the mechanisms of these proteins will contribute to our understanding of not only the toxicity of the venom components, but also the normal physiological events of haemostasis. Studies on the structure and functional relationships are very useful. Such studies will undoubtedly help in the development of different drugs, which might be useful for several medical applications.

This work is a modest attempt by the author towards an understanding the toxic potentials of snake *Naja naja* venom in different tissues of albino rats. This study is far from being comprehensive, yet the author remains hopeful that the present study would contribute useful information to the existing knowledge on changes in metabolic activities in protein metabolism during the effect of snake *Naja naja* venom. The author remains pardonable for any error which may have crept in due to oversight and for any investigative lacunae which are due to limitations in facilities and infrastructure.

This thesis is composed of five chapters. The first chapter deals with the determination of lethal dose, LD_{50} by oral administration of snake *Naja naja* venom. The second chapter deals with alterations in hematological profiles. The third chapter deals with enzymes of protein metabolism and Identification of DNA damage. The fourth chapter deals with detoxification enzymes. The fifth chapter deals with the histopathological changes by light microscopy in different tissues of albino rat.

CONTENTS

General Introduction

Snake bite is a serious problem in rural India. The situation is aggravated by the fact that most rural folk, even today, are superstitious, especially when it comes to the subject of snakes. Many misconceptions and myths about snakes are created by the so called faith healers or 'Ojhas' who monopolize on the rural folk. Until today, the only effective medical treatment for poisonous snake bite is antivenom therapy.

With the passage of time and with the advancement of science, more has been learned of snakes and their venoms, as a result, the fear of snakes has taken a backseat and has opened up a whole new field of snake venom research. Snake venom is regarded as the most complex of all known poisons. The complexity of snake venom can be visualized from the wide variety of pathophysiological conditions observed in snake bite patients. It seems that the snake venom is well equipped with toxins to disturb the whole physiological system of the victim.

SNAKE VENOMICS

There are approximately 2,700 species of snakes, of which only one-fifth are venomous (Mebs, 2002). These toxin containing snakes (superfamily: Colubroidea) can be categorized into four families: Viperidae, Elapidae, Atracta-spididae, and Colubridae (Fox and Serrano, 2008). Each snake family may have similar venom proteins. Nevertheless, amino acid sequences, protein abundances, protein characteristics, and pathological activities of shared venom proteins in each individual family differ. According to the major toxic effect of snake venom in animals, venoms can be classified as hemorrhagic or neurotoxic (Juarez *et al.*, 2004).

Most Viperidae (vipers, pit vipers, and rattle snakes) venoms have little neurotoxic activity but have an abundance of serine proteinases, metallo-proteinases and, C-type lectins activities that hinder the coagulation cascade, the normal hemostatic system and tissue repair. Therefore, envenomation by these snakes generally result in excessive bleeding, lack of clotting, hypo-fibrinogenemia, hypotensive, local tissue necrosis, and inflammatory effects

2

(Braud *et al.,* 2000; Morita, 2004). In contrast, the Elapidae family of snakes (mambas, cobras, coral snakes, and especially Australian snakes) has venoms that generally contain neurotoxins that disable muscle contraction and cause paralysis. Venoms of the Colubridae snakes share some similar activities to both the Viperidae and Elapidae snakes, and the snake venoms of the Atractaspididae family have a variety of peptide toxins that affect the cardiovascular system.

Venom toxins, secreted by a pair of specialized glands connected to the fangs by ducts, are likely to be evolved from proteins with normal physiological functions. These toxins are recruited into the venom proteome before the diversification of the advanced snakes (Morita, 2005; Fry *et al.,* 2006; Mackessy, 2006). Venom proteins have multiple functions, including immobilizing, paralyzing, killing, digesting prey and deterring competitors. Venom represents a key adaptation in ophidian evolution that allows advanced snakes to transition from a mechanical (constriction) to a chemical (venom) means of subduing and digesting large prey. Inspite of the fact that snake venoms contain many distinctive proteins, they belong to only a few families: enzymes (serine proteinases, Zn^{2+} metalloproteinases, phospholipase A2, L-aminoacid oxidase) and proteins without enzymatic activity (C-type lectins, disintegrins, natriuretic peptides, myotoxins, cysteine rich secretory protein [CRISP] toxins, nerve and vascular endothelium growth factors, cystatin, and Kunitz-type proteinase inhibitors).

Snake venom serine proteinases can either trigger or inhibit the specific blood factors, hence affecting platelet aggregation, coagulation, or fibrinolysis (Markland, 1997). Zn^{2+} metalloproteinases degrade extracellular matrix proteins, thereby inducing local hemorrhage, while PLA2 causes severe local swelling followed by necrosis. Snake venom PLA2 isoenzymes are multi-functional enzymes that have many diverse activities such as presynaptic and postsynaptic neurotoxicity, myotoxicity, cardiotoxicity, anticoagulant effects, platelet aggregation (inhibition or initiation), antihaemorrhagic activities, convulsant activities hypotensive activities, edema-inducing activities, and organ or tissue

3

damage activities (Menez, 2002). C-type lectin like proteins (CLP) are multimeric molecules (Fukuda *et al.*, 2000) which include inhibitors and activators of coagulation factors V (AaACP), IX, and X (botrocetin, FIX /X- binding protein). CLP can also compete with von Willebrand factor binding to the platelet membrane. GPIb/IX complex and either block (echicetin, agkicetin, flavocetin, and tokarecetin) or promote (alboaggregins A and B) platelet aggregation. Disintegrins are released from venoms by proteolytic processing of PII Zn^{2+} metalloproteinases and inhibit integrin-ligand interactions (McLane *et al.*, 1998). Each protein family differs from another in protein abundances, sequences, and biological activities (Calvete *et al.*, 2007a; Calvete *et al.*, 2007b; Calvete *et al.*, 2007c; Juarez *et al.*, 2006).

The ophidian evolutionary history retained in the venom composition provided insight into potential taxonomical value. Investigating the characterization of proteins and peptides of snake venom also reveal a number of potential benefits for basic research, clinical diagnosis, development of new research tools and drugs of potential clinical usage, and antivenom production strategies (Menez *et al.*, 2006). Disintegrins are valuable tools for identifying integrin-binding sequence motifs that require selective integrin inhibition. Also, disintegrins have numerous applications for platelet thrombosis, angiogenesis, cancer, bone destruction, and inflammation. Enzymes from cobra venom could be used to treat and prevent Parkinson's and Alzheimers's disease (Barker *et al.*, 2000; Zhao *et al.*, 2001), and the venom from snakes in the viper family has shown to promote tumour reduction (Marcinkiewicz *et al.*, 2003).

COMPOSITION OF SNAKE VENOM

Snake venoms are probably the highest concentrated secretory fluids. The content in solid matter of venoms was found between 16 to 52 % in different species. This is in contrast with mammalian secretory fluids, like human saliva with 0.1 to 1% gastric fluid with 0.5 to 0.8%, or pancreatic juice with 0.4 to 1 %

4

dry matter. The outstandingly high solid concentration is being achieved in snake venoms by mechanisms that are yet fully understood (Stocker, 1990).

The venom composition may vary depending on age, geographic origin, and on an individual level, as concluded from toxicity determinations and measurements of enzyme activities (Momeno *et al.*, 1988). Over 90% of the solid snake venom components are proteins or peptides, and the toxic or biological effects of the venoms are mostly caused by these substances. It is also this class of substances from which products of practical usefulness in medical diagnostic and therapy were developed. The nonprotein fraction of snake venoms consists of inorganic anions and cations, of low molecular substances like amino acids, small peptides, lipids, nucleosides and nucleotides, carbohydrates and amines (Bieber, 1979).

PROTEINS AND POLYPEPTIDES IN SNAKE VENOM

A large number of proteins and polypeptides have been isolated from snake venom, purified to homogeneity and their properties and actions were investigated. The amino acid composition, the primary structure of numerous short and long chain neurotoxins, enzymes and enzyme inhibitors have been determined. The structure of many enzymes have been elucidated either by direct amino acid sequence determination or derived from the base sequence of the gene coding for the particular venom protein (Russel, 1980).These investigations led to the general conclusion that elapid venoms contain high amounts of low molecular peptides and are rather poor in enzymes, where as viperid venoms contain higher molecular proteins and are rich in enzymes (Tu, 1988).

GENERAL CHARACTERIZATION OF VENOM PROTEINS

Snakes and their venoms have fascinated man for several centuries. Tales of venomous snakes and their inherited dangers have passed from generation to generation. Although there have been scattered reports on the topic of snake venom for more than a century, thousands of publications on this topic within the last decades reflect a shift from historical interest to contemporary fascination.

Snake venoms are complex mixtures containing proteins and polypeptides which induce several pharmacological symptoms in the victim. Hundreds of protein toxins have been purified from venoms and the mechanisms by which they induce pharmacological effects have been investigated. These proteins provide useful tools for understanding molecular events in normal physiology. A number of purified toxins from the snake venom have been used as models for analyzing the mechanisms of action of neurotransmitters and their inhibitors. Many toxins from variety of snakes have served as key molecular probes for the characterization and classification of ionic channels and evaluation of their functional roles. Numerous snake venom proteins also interact with components of the human haemostatic system. They interfere in blood coagulation and platelet aggregation. Several venom enzymes have been used clinically as anticoagulants and other venom components are being used in pre-clinical research to examine their possible therapeutic potential. Some snake venom proteins are enzymes, such as phospholipase A2 (PLA2), phosphodiesterase, L-amino acid oxidase, 5'-nucleotidase, thrombin like enzymes (ancrod, atoxin, crotalse, reptilase), and thrombocytin. Such enzymes have been purified from snake venoms and sold commercially and are being used in medicine as therapeutic drugs or in the diagnostic tests. Others serve as tools for the synthesis, transformation and modification of specific chemical compounds.

ENZYMES IN SNAKE VENOMS

Most venoms contain a variety of enzymes and many are hydrolytic with the notable exception of L-amino acid oxidase, which causes the oxidative deamination of aminoacid. The variety of enzymes present in snake venoms is summarized here.

6

A.1 HYDROLYTIC ENZYMES

a. Phospholipase A₂ (Phospholipase A. Lecithinase)

Among the many hydrolytic enzymes isolated from snake venom phospholipase A2 is the most extensively studied (Rosenberg, 1979). Its specificity is directed toward "the site of fatty acids" rather than "the type of fatty acids". Some venoms contain multiforms of phospholipase A2 (isozyme) that have different molecular weights (11,000-15,000 Da), isoelectric points, and immunological properties.

The enzyme itself will cause mild myonecrosis, and the necrotic activity can be greatly enhanced by the addition of phosphatidylcholine. Apparently phospholipase A2 itself is an indirect agent, producing lyso-phosphatidylcholine, which is the direct agent. Phospholipase A2 itself can be separated from the main toxic fraction; however, many presynaptic toxins such as β-bungarotoxin and notexin have weak phospholipase A2 activity (Rosenberg, 1979).

b. Phosphodiesterase

Snake venom commonly contain enzymes that hydrolyze phosphodiester bonds. There are two types of phosphodiesterases one is an exonuclease and the second is an endonuclease.

Exonuclease

Exonuclease removes successive mononucleotide units from the poly-nucleotide chain in stepwise fashion, starting from the 3^1 end. Snake venom exonuclease can hydrolyze almost any type of polynucleotide of any chain length. The type of bases, linkages, and sugars do not have much effect on the rate of hydrolysis. Thus, the enzyme hydrolyzes RNA, DNA, synthetic poly-nucleotides, and native or denatured DNA.

7

Endonuclease

Some snake venoms do possess endopolynucleotide activities and can hydrolyse RNA and DNA, producing oligonucleotide fragments (Elliott, 1978).

c. Phosphomonoesterase

Snake venoms contain nonspecific as well as specific phosphomono-esterases. Frequently it is referred as phosphatase, depending on its optimum pH, by which the enzyme is designated as an acid phosphatase or an alkaline phosphatase.

Nonspecific Phosphomonoesterase

The presence of non- phosphomonoesterase in snake venom is common. This enzyme hydrolyzes a variety of compounds containing phosphomonoester bond. Some venoms contain both acid and alkaline phosphatase, whereas others contain only one type.

Specific Phosphomonoesterase

A specific phosphomonoesterase that is found commonly in snake venoms is 5^1-nucleotide, which hydrolyses a variety of 5^1-nucleotide phosphates. Thus the enzyme does not hydrolyses 2^1 or 3^1-phosphates of a guanosine, adenosine, cytidine, uridine, or adenoside 21, 3^1 – cyclic phosphates. Nucleoside diphosphates or triphosphates are also not hydrolyzed by 5^1-nucleotidase (Iwanaga, 1979).

d. Acetyl cholinesterase

Acetyl cholinesterase is commonly present in the venoms of Elapidae and Hydrophidae but not in those of Viperidae and Crotalidae. Since acetyl cholinesterase is involved in nerve transmission, it was thought at one time that venom acetyl cholinesterase is responsible for the neurotoxic action. Acetyl cholinesterase from snake venom is being used as a model for its nerve and muscle counterpart. They offer an exceptional system for analyzing the mechanism of peripheral site inhibition, because of their wide range of activities (Cousin, 1999).

8

e. Proteolytic enzymes

Certain snake venoms are rich in proteolytic enzymes, especially those of Crotalidae and Viperidae, which have strong endopeptidases. Elapidae venoms usually do not contain endopeptidase but are rich in di and tripeptidases. Some snake venoms contain very specific proteases, cleaving only specific peptide bonds (Tu, 1988).

Endopeptidases

Earlier, venom endopeptidase was considered to be identical to pancreatic trypsin. A number of venom protease were isolated and their specificities were investigated. Unlike trypsin, the sites of hydrolysis of venom proteases are more varied and cannot be generalized. Some venom proteases are metalloproteins that contain calcium and zinc ions.

Specific proteases

There are many varieties of special venom proteases. For instance, some venoms contain the bradykinin- releasing factor (kininogenase), which hydrolyzes two specific peptide bonds in bradykininogen to release bradykinin. Snake venom are also known to have a pronounced effect on the blood coagulation system. The factors affecting the blood coagulation are also proteases. Venom collagenase hydrolyzes collagen but not other proteins; it is believed that venom collagenolytic enzyme is a true collagenase. Collagenase activity is strongest in the venom of Crotalidae and less potent in the venom of viperidae. Most Elapidae venoms contain no or minimal collagenase activity (Markland, 1983).

f. Arginase esterase and other esterases

Arginine esters are convenient substrates for trypsin assay because the enzyme can be measured spectrophotometrically. Like trypsin, some snake venoms hydrolyze this substrate. Therefore, it was thought that snake venoms contain trypsin or trypsin like enzymes, however, venom proteases and trypsin have different sites of proteolysis. The protease and arginine ester hydrolase were

9

eventually separated into different fractions. Arginine esterase activities are usually associated with very specific proteases such as bradykinin-releasing enzyme and with blood coagulation activities. Usually arginine ester hydrolases are present in the venoms of Crotalidae and Viperidae, but they are not found in the venoms of Elapidae and Hydrophiidae. Thus the enzyme distribution in snake venoms has taxonomic significance at the family level (Tu, 1988; Geiger *et al.,* 1977).

g. Hyaluronidase

Hyaluronidase is commonly present in venoms of Elapidae, Viperidae, and Crotalidae. Hyaluronic acid, a mucopolysaccharide present in the skin, connective tissue, and bone joints, serves to promote intercellular adhesion or acts as a lubricant. Chemically, hyaluronic acid consists of repeating units of the type $(-N-G)_n$ where N is N- acetyl-d-glucosamine and G is d-glucoronic acid. The glycosidic linkages N to G and G to N are β-1, 4-and β-1,3-respectively. Snake venom hyaluronidase hydrolyzes the glucosaminidic bond between C_1 of the glucosamide unit and C4 of glucuronic acid.

The enzyme is frequently referred to as the spreading factor because hydrolysis of hyaluronic acid facilitates toxin diffusion into the tissues of the victim. Hyaluronidase itself can be separated from the toxin fraction, hence is not a main factor (Tu, 1988).

h. NAD nucleosidase

NAD nucleosidase or some times simply called NADase is present in some snake venoms. It hydrolyzes the nicotinamide N-ribosidic linkage of NAD. The products of this reaction are nicotinamide and adenosine diphosphate ribose.

10

A2. NON HYDROLYTIC ENZYMES

L-amino acid oxidase a nonhydrolytic enzymes is widely distributed in snake venoms. It is largely responsible for the yellow color of the snake venoms. There are many isoenzymes of L-amino acid oxidase with different isoelectric points. The purified enzyme is a homodimeric glycoprotein of 110,000-140,000 Da and contains FAD/ or FMN as its prosthetic groups for each monomer. They also differ in their substrate specificities and stability in different storage conditions. The enzyme is capable of generating H_2O_2 by catalyzing oxidative deamination of L-amino acids (Ueda *et al.*, 1988).

L-Amino acid +2FMN (2FAD) $\longrightarrow$ FMNH$_2$ (FADH$_2$) + α KA +NH$_3$

FMNH$_2$ + O$_2$ + H$_2$o $\longrightarrow$ H$_2$O$_2$ + FMN

L-amino acid oxidase from "King Cobra" *(Ophiophagus hannah)* venom inhibits platelet aggregation (Li *et al.*, 1994) and the same enzyme from "Crotalus atrox venom" has been identified as a major hemorrhagic factor (Braganca *et al.*, 1970). The cytotoxicity of L-amino acid oxidase from "Crotalus atrox venom" has been observed in number of cancerous cell lines and in normal proliferating cells. L-amino acid oxidase from "western diamond back rattle snake" induces apoptosis in specific cell lines (Eldadah *et al.*, 1996).

PROTEINS ACTING ON THE BLOOD COAGULATION

Snake venoms exert profound effect on the blood coagulation system; some accelerate the process itself and others retard it. Frequently, snake venoms are divided into two types, coagulant (procoagulant) and anticoagulant, but it is not unusual to find one venom containing both coagulant and anticoagulant factors, however a venom exerts coagulant or anticoagulant effects depending on the concentration used (Marsh, 1994).

11

TOXIC COMPONENTS OF SNAKE VENOM

a. Neurotoxins

The venoms of many snake species contain agents that affect nerve functions of the prey animal, causing cramps, convulsions, or paralysis. These agents, according to their mode of action, may be subdivided into a class of nonenzymatic, postsynaptic neurotoxins or α-neurotoxins which block neurotransmitter receptors (Dolly *et al.,* 1986).

To date, over 100 post synaptic neurotoxins have been isolated and the amino acid sequence of them has been elucidated. The postsynaptic neurotoxins may be allocated either to the class of short chain neurotoxins consisting of 60 to 62 amino acid residues cross linked by 4 disulfide bridges or to the class of long chain neurotoxins containing 66 to 74 amino acid residues and 4 to 5 disulfide linkages. The presynaptic neurotoxins are mostly toxic phospholipases A_2 and they exert the catalytic function of this type of enzyme.

b. Hemotoxins

Purified snake venom proteins have become valuable tools in basic research and in hemostaseology. "Procoagulant" as well as "anticoagulant" venom components have been identified in *in vitro* tests. Smaller doses of procoagulant venom components applied to large organisms as in the case of snake-bite accidents in humans may cause a consumption coagulopathy with localized or generalized bleeding. Highly purified, specific fibrinogen coagulant venom proteinases are used in human medicine to produce therapeutic defibrinogenation (Meier and Stocker, 1991).

Venom components affecting platelet functions, blood coagulation, or fibrinolysis, as well as venom enzymes causing vasodilation or increased capillary permeability, may contribute in a synergetic manner to this hemorrhagic action. However, the presence of venom components that directly damage the blood vessel wall is the major cause of such bleeding. Venom proteins with a direct

12

damaging effect on the vessel wall are designated as hemorrhagic principles (Stoker, 1990).

c. Myotoxins

Snake venom polypeptides that induce skeletal muscle contraction or produce local myonecrosis or myoglobinuria are categorized as myotoxins. Samejima *et al.*, (1988) found in *Crotalus adamanteus* venom the non enzymatic myotoxin CAM which was identified as a dimeric, basic peptide composed of two subunits with a molecular weight of 11,000 Da (Samejima *et al.*, 1988).

A basic, dimeric polypeptide with a subunit molecular weight of 16,000 Da isolated from the venom of the snake *Bothrops mummifer* caused muscle cell damage *in vivo* and *in vitro*, and upon i.m. injection into mice, lead to the release of Creatine kinase. This myotoxin appeared devoid of phospholipse A_2 activity (Gutierrez *et al.,* 1986). A myotoxin devoid of phospholipase A_2 activity was isolated from *Bothrops jararacussi* snake venom. The myotoxin, called Bothropstoxin or B^{th} TX, was characterized as a single chain peptide with a molecular weight of 13,000 Da, bearing 16 half cysteine residues. Its isoelectric point was 8.2 (Homsi-Brandeburgo *et al.,* 1988). In addition to these nonenzymatic myotoxins, venom phospholipases A_2 as well as hemorrhagic metalloproteinase's may also exert muscle cell damaging effects (Stocker, 1990).

d. Cytotoxins

Cytotoxins are toxic polypeptides consisting of 60 to 62 amino acid residues with four intramolecular disulfide bonds. The pharmacological actions of cytotoxins comprise haemolysis, cytolysis, depolarization of muscle membrane, and specific cardiotoxicity.

The mode of action of these nonenzymatic cytolytic agents probably comprises the activation of cellular phospholipases A_2 as observed with the "direct lytic factors" of the venoms of *Hemachatus hemachatus* and of *Naja naja atra*

(Shier, 1980).The *in vitro* cytotoxic substance from *Agkistrodon contortix* laticintus and King Cobra (*Ophiophagus hannah*) snake venoms has been indentified as L-amino acid oxidase and the cytotoxic mechanism was due to the generation of hydrogen peroxide from oxidative deamination of L-amino acids catalyzed by enzyme (Ahn *et al.,* 1997; Souza *et al.,* 1999).

NON PROTEINOUS SNAKE VENOM COMPONENTS

Since the toxicity of snake venoms definitely relates to proteins or polypeptides and these constituents represent the major fraction of the total venom, relatively few investigations have been performed concerning the nonprotein organic components such as lipids, carbohydrates, riboflavin nucleosides, nucleotides, amino acids, biogenic amines and inorganic snake venom components such as cations and anions.

PRIMARY TREATMENT OF SNAKE BITE

Reassurance of the patient is important and if available, aspirin or alcohol in moderation is helpful for their calming effects. The site of the bite should be wiped out, not incised, because incision can aggrevate bleeding, especially in bites causing non-clotting blood, damage nerves and tendons, introduce infection and delay healing. A tourniquet may be applied on the bitten limb, a few inches above the bite site; however the tourniquet should not be applied very tightly. Though the efficacy of the tourniquet in delaying spread of venom is doubtful, it will nevertheless give the patient some physiological boost. Application of a crepe bandage over the bite site is beneficial.

SCOPE OF THE PRESENT STUDY

Snake bite is one of the most neglected public health issues in poor rural communities living in the tropics. Because of serious misreporting, the true worldwide burden of snake bite is not known. South Asia is the world's most heavily affected region, due to its high population density, widespread agricultural activities, numerous venomous snake species and lack of functional snake bite control programs. Despite increasing knowledge of 'snake venom' composition and mode of action, good understanding of clinical features of envenoming and sufficient production of antivenom by Indian manufacturers, snake bite management remains unsatisfactory in this region.

Field diagnostic tests for snake species identification do not exist and treatment mainly relies on the administration of antivenoms that do not cover all of the important venomous snakes of the region. Poorly informed rural populations often apply inappropriate first-aid measures and vital time is lost before the victim is transported to a treatment centre, where cost of treatment can constitute an additional hurdle. The deficiency of snake bite management in South Asia is multi-causal and requires joint collaborative efforts from researchers, antivenom manufacturers, policy makers, public health authorities and international funders.

Since ancient times, snakes have been worshipped, feared in South Asia. Cobras appear in many tales and myths and are regarded as sacred by both Hindus and Buddhists. Unfortunately, snakes remain a painful reality in the daily life of millions of villagers in this region. Indeed, although antivenom is produced in sufficient quantities by several public and private manufacturers, most snake bite victims don't have access to quality care, and in many countries, both morbidity and mortality due to snake bites are high. The neglected status of snake bite envenoming has recently been challenged apart from the production of antivenom, snake bite envenoming in South Asia shares few of the characteristics of a neglected area. This review aims at summarizing and discussing the potentiality of snake *Naja naja* venom on different target organs, different biochemical aspects, histopathological conditions. Undoubtedly many interesting aspects and physiological actions to be discovered in the toxic secretions of snake *Naja naja* venom in the present research work.

Materials and Methods

EXPERIMENTAL DESIGN

Species: Albino rat

Venom selected: Snake *Naja naja* venom, Lyophilized powder was obtained from Irula Snake Catchers, Industrial co-operative society, Vadanamelli, Perur post, Kanchipuram District, Tamilnadu.

Concentration selected: Fiftieth fold (1/50th), lower concentration of LD_{50} was selected for sub lethal treatment to the experimental rats.

Course of study: 24 hrs, 48 hrs, and 72 hrs with 24 hrs time interval in each group.

Route of administration: Oral

Tissues selected: Liver, Kidney, Brain, Heart and Blood.

Snake *Naja naja* stock solution: Stock solution of snake *Naja naja* venom was prepared in 1% Sodium chloride. Snake *Naja naja* venom test solution was prepared by diluting the stock solution with 1% Sodium chloride.

Selection of sublethal treatment to the experimental model: The acute oral LD_{50} value of snake *Naja naja* venom was determined, Fiftieth fold lower (1/50th) concentration was selected as sub lethal dose to study the effect of snake *Naja naja* venom. Healthy adult albino rats of same age (100±10 days) and weight (150±10 g) were divided into four groups having ten animals each. First group was control, second group was 24 hrs, third group was 48 hrs, fourth group was 72 hrs. The 24 hrs, 48 hrs and 72 hrs groups were termed as experimental animal groups or envenomated groups. To the animals of second group, i.e., 24 hrs, snake *Naja naja* venom (1st day) was administered orally by gavage method. To the third group of animals i.e., 48 hrs, snake *Naja naja* venom was administered (i.e. 1st and 2nd day). To the fourth group of animals i.e., 72 hrs, snake *Naja naja* venom was administered (1st, 2nd day and 3rd day).

17

Isolation of tissues

The control and experimental animals after the stipulated time (i.e., 24hrs, 48 hrs and 72 hrs) were sacrificed and the tissues were isolated, cleaned in physiological saline and processed immediately for microscopic analysis. The tissues were also quickly isolated under ice cold conditions and stored in deep freezer at -80°C for biochemical analysis.

The experimental design has the following parameters for investigation.

Selection of experimental model:

The present study was carried out in the Albino rats. This animal was chosen for the study because of the following reasons.

1. Albino rats are considered as an animal model for experimental studies.

2. Albino rats are relatively small, can be handled easily and require less feed.

3. The rats are preferred because of the plethora of toxicity data that exists for rodents (Bruce, 1985).

4. The laboratory Albino rats are close to the field rats in its physiology, hence, these animals were selected.

5. Extrapolation of the rats to the human being (man) is usually done for risk assessment (Moser, 1990) for a wide variety of venoms (Groten *et al.,* 1997).

6. Acute toxicity test was conducted as it throws light on how the animal responds to a 24hrs, 48 hrs, 72 hrs of time interval in snake *Naja naja* venom.

7. To study the effect of snake *Naja naja* venom in the Albino rats.

8. To identify the effects of snake *Naja naja* venom on the potential

18

target tissues namely liver, kidney, brain, heart and blood are selected for different parameters.

9. The choice of the animal model to be used in toxicity studies appears critical as different tissues may be the target of the drug's toxic effect in different animal species (Roncaglioni *et al.*, 1982).

Procurement of experimental animals

Healthy wistar strain Albino rats of the same age group 100±10 days and weight 150±10 grams were selected as experimental animals for the present study. The rats were collected from Indian Institute of Science (I.I.Sc.), Bangalore. Prior to experimentation the animals were acclimatized according to the instructions given by Behringer (1973). Rat feed was supplied by Sai Durga feeds and foods, Bangalore.

Maintenance of animals

The rats were maintained at laboratory conditions in the animal house at $25\pm2^{\circ}C$ with a photoperiod of 12hrs light and 12hrs darkness throughout the course of the present study. The rats were fed with standard pellet diet and water *ad libitum*.

Venom selected: Snake *Naja naja* venom

Highly poisonous snake *Naja naja* venom was selected for the present investigation. Crude lyophilized powder venom snake *Naja naja* venom was obtained from Irula Snake Catchers, Industrial co-operative society, Vadanamelli, Perur post, Kanchipuram District, Tamilnadu, India.

The physical and chemical characteristics of Snake *Naja naja* venom

The following are the specifications of snake *Naja naja* venom used in the present study.

Generic name   :   *Naja naja*

Chemical name	:	*Naja naja oxiana*
Color	:	Pale yellow color
Synonym(s)	:	*Naja naja atra cardiotoxin*
Specific gravity	:	1.03 to 1.07
Compound descriptor	:	Natural product
Species observed	:	Albino rat
Route of exposure	:	Oral
Storage temperature	:	-20^0C
Composition protein	:	35% biuret
CAS (Chemical Abstracts Services) registry	:	13146-36-6
MSDS	:	Venom, snake, *Naja naja oxiana*
CAS No	:	11119-60-1
Physical state	:	viscous fluid
CBN number	:	CB1990898
Odor	:	Odourless and tasteless
Solubility	:	Sodium chloride
Specific gravity	:	1.03 – 1.07
Organic solvent (s)	:	BSA

CHAPTER SCHEME: Thesis consists of 5 chapters

CHAPTER -I

TOXICITY EVALUATION OF SNAKE *NAJA NAJA* VENOM

Toxicity evaluation was estimated by Finney (1971).

Lethal dose (LD_{50}) of snake *Naja naja* venom was determined by "Probit method" of Finney (1971). Dose and mortality were noted and a graph was plotted between snake *Naja naja* venom concentration and probit kill. LD_{50} was the dose at which 50% of the test animals were killed.

CHAPTER –II: HAEMATOLOGY

HAEMOGRAM

2.1 Red blood corpuscle (RBC) count

RBC count was made with a Neubauer crystalline counting chamber as described by Samuel (1977).

The blood of the animal was collected in a vial containing 2% ethylene diamine tetra acetic acid (EDTA) as an anticoagulant. The blood was drawn up to 0.5 mark in RBC pipette and immediately the diluting fluid was drawn up to the mark 101 (thus the dilution was 1:200). The solution was mixed well by shaking gently. The solution was allowed to stand for 2 or 3 minutes. The counting chamber and cover glass were cleaned and the cover glass was placed over the ruled area. Again the solution was mixed gently and stem full of solution was expelled and a drop of fluid was allowed to flow under the cover slip holding the pipette at an angle of 40^0, it was allowed to stand for 2 to 3 minutes to allow RBC to settle. Afterwards the ruled area of the counting chamber was focused under the microscope and the number of RBC's were counted in five small squares of the RBC column under high power and the number of RBC per cu mm were calculated accordingly.

$$\frac{\text{Number of cells} \times \text{dilution factor} \times \text{depth factor}}{\text{Area counted}}$$

Estimation of haemoglobin concentration (Hb)

The haemoglobin concentration was estimated by Acid - haematin method by Samuel (1977).

Blood was collected directly from the eyeball up to 20 cu mm in the Hb pipette and the outer side was wiped out. This was transferred into the graduated tube containing N/10 hydrochloric acid. N/10 hydrochloric acid was taken up to 20 mark in a graduated tube. The pipette was rinsed two or three times with dilute hydrochloric acid. It was allowed to stand for 10 to 20 minutes after thorough mixing. Then N/10 HCl was added drop by drop, mixing between each addition until the blood color matched with the standard color. The results were read from the scale on the graduated tube and the Hb concentration was expressed in grams percent.

Estimation of packed cell volume (PCV)

PCV was estimated by Micro hematocrit method (Samuel, 1977). The blood of the animal was drawn into ependoff tubes containing the anticoagulant, by capillary action to 2/3 of their length. The tubes were tapped to permit blood to flow towards end and to provide sufficient space to prevent outflow when the opposite ends were sealed. The outside of the ependoff tubes were wiped free of blood and the index finger was placed over the moist ends to hold the column of the blood in place as the opposite dry ends were forced into the sealing material to form a tight plug. The ependoff tubes was placed in the centrifuge with the sealed ends pointing outward and centrifuged at 12,000 rpm for 5 minutes. The optical density was read on the reader which gives the direct haemotocrit value in volumes percent.

22

Mean corpuscular volume (MCV)

Mean Corpuscular Volume was estimated by Samuel (1977).

MCV expresses the average volume of the red blood cells. For obtaining the mean corpuscular volume, the packed cell volume is divided by red blood cell count and the result was multiplied by 10. MCV was expressed in cubic microns (cu μ).

To determine the average volume of a single red cell in cubic microns.

$$M.C.V : \frac{\text{Packed cell volume} \times 10}{\text{Red Blood Cells in millions per c.mm}}$$

Mean corpuscular haemoglobin (MCH)

Mean Corpuscular Haemoglobin was estimated by Samuel (1977).

MCH represents the average weight of haemoglobin contained in each cell. MCH was influenced by the size of the cell and concentration of haemoglobin. For getting MCH the Hb concentration was usually divided by red blood cell count and the result was multiplied by 10 and was expressed as picograms (pg).

To determine average haemoglobin content of a single red cell in micro-micrograms.

$$M.C.H : \frac{\text{Haemoglobin} \times 10}{\text{RBCs in millions per c.mm}}$$

Mean corpuscular haemoglobin concentration (MCHC)

Mean Corpuscular Haemoglobin Concentration was estimated by Samuel (1977).

MCHC refers to the average concentration of the Hb in the red blood cells. In contrast to MCH, MCHC is not influenced by the size of the cell. For getting MCHC the haemoglobin was divided by packed cell volume and the result is multiplied by 100. The MCHC value was expressed in terms of percentage.

23

$$M.C.H.C\% : \frac{Haemoglobin \times 100}{Packed\ cell\ volume}$$

Platelet (Thrombocytes) count

Platelet count was estimated by a Direct method, Samuel (1977).

The blood was drawn into a test tube and aspirated a little of Platelet diluting fluid into the RBC pipette and expel the fluid. The blood was diluted in 1 in 200 with diluting fluid. The counting chamber was kept in a Petridish containing wet cotton wool and was waited for 15 minutes, until the platelets have settled properly. The platelets were counted in a whole finely ruled area (red cells) using high power objective. The Platelets were liliac colored and was 1/7 to 1/2 to the diameter of RBC and usually oval, rod or comma shaped.

$$Platelets\ per\ cu.mm\ of\ blood = No.of\ Plates \times \frac{RBC'S}{1000}$$

White blood corpuscles (WBC) count

WBC count was estimated by Samuel (1977).

The blood was drawn from the animal into a vial and the blood was taken into a WBC pipette up to 0.5 mark and immediately the diluting fluid was drawn up to 11 mark. The solution of the blood along with the WBC diluting fluid was mixed thoroughly by shaking gently. Allow the cells for 3 minutes to settle. The Neubauer counting chamber is used to count the cells in the four corners blocks. Each of these 4 square millimeter area is sub-divided into 16 squares, by using the low power objective and the medium ocular. The difference between the two square millimeter areas should not be more than 10 WBC's. The WBC count was expressed in cu mm.

24

2.2 Haemogram of differential leukocyte count

Differential leukocyte count was estimated by Samuel (1977)

A drop of the animal blood was taken and placed on a clean glass slide for about 1-2 cm. The blood on the slide was spreaded from one end with the help of a spreading slide placed at an angle of 45° approximately. The slide was placed flat on two glass rods over a sink and was covered with leishman stain. The stain was diluted drop by drop addition of buffered water and stained for a period of 5-7 minutes. The stain was drained and washed with water and air dried and observed under microscope. The cells were counted under high power oil immersion objective from the edge of the smear moving the smear towards center. The Leucocytes were identified and the movement was repeated till a total 100 cells were counted. The different morphological types of leukocytes such as neutronphils, lymphocytes, monocytes, eosinophils, basophiles were observed and these different types of leukocytes were expressed in the percentage.

2.3 HAEMOGRAM OF SERUM CREATININE AND LIPID PROFILE

Serum creatinine

Serum creatinine levels were estimated by the method of Jaffe's reaction (Folin and Wu, 1991).

Preparation of protein free filtrate

The test tubes were taken and 2 ml of serum is added accurately followed by 4 ml of the distilled water. 1ml of 10% sodium tungstate and 1 ml of 2/3 N sulphuric acid was taken and mixed well after each addition. Two test tubes were taken and labeled as 'Blank and Test'. To the blank test tube 4 ml of the distilled water and 2ml of alkaline picrate solution (prepared freshly by mixing 5 volumes of saturated picric acid with 1 volume of 10% sodium hydroxide) were added. To the test tube, 4ml of protein free serum filtrate, and add 2ml of Alkaline picrate solution (prepared freshly by mixing 5 volumes of saturated picric acid with 1 volume of 10% sodium hydroxide) were added. The contents of the two test tubes were mixed properly and allowed to stand for 15 minutes. The optical density was measured at 520 nm against blank set at zero.

   i.   Alkaline picrate solution was prepared freshly before use.

   ii. If color was too deep, 1 ml of the filtrate was taken and made up to 4ml with distilled water.

Serum cholesterol

Serum cholesterol levels were estimated by the modified Liebermann Burchard method (King and Wootton, 1959).

The test tubes were taken and 0.2 ml of the serum and 10 ml of ethanol-acetone solvent were added. The test tubes were kept in a boiling water bath (with shaking) until the ethanol-acetone solvent begins to boil. The test tubes were taken out of the water bath and shaken well continuously for 5 minutes. The addition of the ethanol-acetone solvent causes proteins to precipitate. The test tubes were cooled to room temperature and the contents in the test tubes were taken in a centrifuge tube and centrifuged at 2000 rpm for 5 minutes. The supernatant was collected in a fresh test tube and was kept in a boiling water bath until the supernatant i.e. solvent gets evaporated. The pellet was redissolved in 2 ml of chloroform. The optical density was measured at the 640 nm against blank set at zero.

Triglycerides

Triglycerides were estimated by Samuel (1977).

0.5ml of the test sample, 0.5 ml of the distilled water, 0.5ml of the standard solution were added to 3 test tubes respectively. 3.5 ml of isopropanol, 1ml of 0.08N H_2So_4, 2ml of heptanes were added to three test tubes respectively, and the contents were mixed by vortexing. Layers were allowed to separate. From these three layers 0.2ml of the aliquots were taken from each tube by micropipette into the test tubes stoppered with Teflon-lined screw caps and labeled the three test

26

tubes as test, blank, and standard. 3 ml of sodium methylate solution was added and mixed with the help of vortex mixer. Test tubes were incubated at 60°c for 15 minutes. 1 ml of sodium metaperiodate solution was added and mixed well and then 1 ml of acetylacetone reagent was added. The contents were mixed well and was boiled in a boiling water bath at 60°c for 10 minute. The test tubes were cooled at room temperature and then centrifuged at 2000 rpm for 10 minutes. The supernatant was taken and the optical density for both the test solution and the standard solution was read at 410 nm, setting the absorbance to zero using the reagent blank. The triglycerides were calculated by using the formulae.

The triglyceride concentration was mg/dl as follows:

The triglyceride concentration in (mg/dl) = A1/A3 × 300.

2.4 HAEMOGRAM OF LIVER FUNCTION TESTS

Serum proteins

Total proteins levels were estimated by the specific gravity method modified by Copper sulphate method (Wootton, 1974).

The serum was separated from the blood and copper sulphate solution was added. The specific gravity of the solution in which the raise or fall in the drops of the serum was noted. The fragments of the drops remaining on the surface of the solution should be sunk with a glass rod before adding the next drop. The solution was changed its color after 25 drops per 25ml of copper sulphate solution.

Alkaline phosphatase (ALP)

Alkaline phosphatase activity was estimated by micro haematocrit method (Wootton, 1964).

The test tubes were taken and 1 ml of the buffer and 1ml of Phenyl phosphate substrate were added and kept in a water bath at 37^0C for 3 minutes. The test tubes were taken out from the water bath and 0.1ml of serum was added and mixed gently and incubated for 15 minutes and the reaction could be stopped

27

by adding 0.8 ml of 0.5N Sodium hydroxide. To the control test tube, 1ml of the buffer and 1ml of the substrate and 0.8ml of 0.5N Sodium hydroxide and 0.1ml of serum were added and mixed properly. To the standard test tube, 1.1ml of the buffer and 1ml of Phenol standard (1mg per 100ml) and 0.8ml of 0.5N Sodium hydroxide were added and mixed properly. To the blank test tube, 1.1 ml of the buffer and 1ml of the distilled water and 0.8 ml of 0.5N sodium hydroxide were added and mixed properly. To all the three test tubes, 1.2ml of 0.5N Sodium bicarbonate followed by 1ml of 4-Aminoantipyrine solution and 1ml of Potassium ferricyanide solution were added and mixed well after each addition. The pH was adjusted till the color was developed. The optical density was read at 510 nm against blank set at zero.

Gamma glutamyl transferase (G.G.T)

Gamma glutamyl transferase levels were estimated by Samuel (1977).

The blood was collected from the animal into a plain centrifuge tube, and was allowed to clot for 1hr. at room temperature ($25^0C \pm 2$) and the serum was collected by centrifugation. 1 ml of the buffer and 50 µl of the sample were taken into a eppendoff tube at 37°C. 100 µl of the gamma glutamyl-p-nitroanilide solution was mixed and the optical density was read at 405 nm against the blank set at zero.

2.5 HAEMOGRAM OF ELECTROLYTES

Sodium

Serum sodium levels were estimated by the Trinder (1915).

8 g of the Uranyl acetate and 30 g of the magnesium acetate was dissolved in 150 ml of distilled waster and 30 ml of glacial acetic acid was added and boiled for 2 minutes and cooled the test tubes and made the volume to 200ml with distilled water and transferred to a 1 litre volumetric flask. The Solution was transferred to a 1litre volumetric flask and made up the mark with absolute alcohol. The solution was stored in a amber colored bottle and 1 ml of

28

1% sodium chloride was added and mixed the contents properly and was allowed to stand for several days until the precipitate of sodium, magnesium and uranyl acetate were formed. The supernatant was used and allowed the solution to stand for 5 minutes. The solution was mixed well for 30 seconds and centrifuged at 2000 rpm for 1 minute. 2ml of the supernatant was taken in a cuvette and the optical density was read at 480 nm against blank set at zero.

Calculation:

$$\text{Serum Sodium} = \frac{\text{Density of unknown}}{\text{Density of the standard}} \times 300\text{mg per } 100\text{ml.}$$

Potassium

Serum potassium levels were estimated by the Kramer and Tisdall (1921).

1ml of the rat serum was pipetted out into a 15ml graduated centrifuge tube and 2ml of Sodium cobalt nitrite solution was added and mixed thoroughly and allowed to stand for 45 minutes and 2ml of distilled water was added and mixed the contents and centrifuged at 1400 rpm for 30 minutes. 0.3ml of the supernatant was taken, and care was taken not to disturb the sediment and 5ml of distilled water was added and recentrifuged at 1400 rpm for 5 minutes. The procedure was repeated 3 times the supernatant fluid in the last washing is removed and excess of 4N sulphuric acid and 1ml of sodium nitrite was added and mixed thoroughly and heated in a boiling water bath at 37°c for 10 minutes until no further color change was observed. 2ml of potassium permanganate solution was added to decolorize the mixture completely and excess of oxalate was determined by titrating the solution with potassium permanganate solution to get a definite pink color.

Serum calcium

Serum calcium levels were estimated by the Kramer and Tisdall (1921).

Two test tubes were taken, one for test solution and the other test tube for the standard solution. 2ml of rat serum and 2ml of distilled water and 1ml of the ammonium oxalate saturated aqueous were added to the test solution and allowed to stand for 30 to 40 minutes. Solution was centrifuged at 5000 rpm for 5 minutes and the supernatant was removed. 4ml of ammonium hydroxide and 2% aqueous solution were added to the test tubes. Solution was mixed well and recentrifuged at 5000 rpm for 5 minutes. Supernatant was removed by repeated washing. 2ml of sulphuric acid was added and the test tube was placed in a boiling water bath at 37°c for 1 minute and the test tube was cooled and was titrated with potassium permanganate using a micro burette until a faint trace of pink persists. The optimum temperature per titration was maintained at 75^0c. Volume of potassium permanganate was recorded. 0.1 ml of the ammonium hydroxide and 2 ml of sulphuric acid was added to the standard solution and was heated in a boiling water bath for 1 minute. 2 ml of the potassium permanganate was added to the blank test tube.

Reading of the unknown =

Reading of blank × 10 mg of the calcium per 100 ml of the serum.

Serum phosphorous

Serum phosphorous levels were estimated by the Gomorri and Bab (1942).

0.8 ml of serum was taken in a test tube and 7.2 ml of 10 % trichloroacetic acid was added and mixed well and it was filtered. Four test tubes were taken, each containing 5 ml, (0.5 ml of the serum) of the filtrate (unknown) and 0.5 ml of the standard (0.025 mg phosphorous) and 4.5 ml of 10 % Trichloroacetic acid (the standard) and 5 ml of the trichloroacetic acid (the blank) were added. 1 ml of Ammonium molybdate and 1 ml of metol solution was added to each test tubes.

The solution is allowed to stand for 30 minutes and the optical density was read at 680 nm against the blank set at zero.

Calculation: Inorganic phosphorous per 100 ml of the serum

$$= \frac{\text{Reading of unknown}}{\text{Reading of the standard}} \times 0.025 \times \frac{1000}{0.5}$$

Serum amylase

Serum amylase activity levels were estimated by the Samuel (1977).

The test solution was taken in a test tube and the serum was diluted with 0.9% saline. 1 ml of starch substrate is pipetted out into a test tube (T) and was placed in a water bath at 37° C. After 3 minutes 0.1 ml of diluted serum was added and mixed and incubated at 37° C for 15 minutes and removed the test tube from the water bath and quickly 8.5 ml of distilled water was added to stop the reaction. 0.4 ml of dilute iodine solution of 0.01N was added. To the control test tube, 1 ml of starch substrate, 8.6 ml of the distilled water, 0.4 ml of diluted iodine solution were added. The colors were compared and the optical density was read at 600 nm against the blank set at zero.

Calculation : The amount of starch digested in the test is therefore

$$\frac{\text{Control} - \text{test}}{\text{Control}} \times 0.4 \, \text{mg}.$$

Lactate dehydrogenase (LDH)

Lactate dehydrogenase activity levels were estimated by Wootton (1974).

1 ml of the buffered substrate was taken in a test tube and 0.1 ml of the serum was added and placed the test tube in the water bath at $25 \pm 0.2°C$ for 10minutes. The reaction was started by adding 0.1ml of NADPH solution and incubated the solution for exactly 15 minutes and removed the test tube from the water bath and immediately 1 ml of Dinitrophenyl hydrazine solution was added. In a control test tube, 1 ml of the substrate and 0.2 ml of the buffer and 1 ml of

31

dinitrophenyl hydrazine were added. To the blank test tube 1.2 ml of the buffer and 1 ml of the dinitro phenyl hydrazine solution were added. All the test tubes were allowed to stand at room temperature for 20 minutes and 10 ml of 0.4 N sodium hydroxide was added to each test tube and mixed properly for 10 minutes. The optical density was read at 510 nm against the blank set at zero.

3A. PROTEIN METABOLISM

3.1 Estimation of total proteins

The total protein content was estimated by the method of Lowry *et al.,* (1951).

2% homogenate of the protein solution was prepared in 10 % TCA and centrifuged at 1000xg for 15 minutes. The supernatant was discarded and the residue was dissolved in a known amount of 1N sodium hydroxide. From this 0.2ml of the residue was taken and to this 4 ml of alkaline copper reagent and 0.4ml of folin phenol reagent (1:1 folin phenol: distilled water) were added. The contents were allowed to stand for 30 minutes at room temperature and the optical density was read at 600 nm in a spectrophotometer against a reagent blank.

The amount of total proteins present in the sample was calculated by using bovine albumin standard and the values were expressed as mg/g wet weight of tissue.

3.2 Estimation of Free amino acids (FAA)

Free amino acid content was estimated by the method of Moore and Stein (1954) as described by Colowick and Kaplan (1957).

5% homogenates of different tissues were prepared in 10% TCA and centrifuged for 15 minutes at 1000xg. To 0.25ml of the supernatant, 2ml of ninhydrin reagent was added and kept in boiling water bath for 6.5 minutes and then cooled. The contents were made up to 10ml with distilled water. The intensity

32

of the color developed was read at 570 nm in a spectrophotometer against a reagent blank.

The total free amino acid content was expressed as μ moles of tyrosine equivalents /g wet weight of the tissue.

3.3 Estimation of protease activity

Protease activity was estimated by the method of Moore and Stein (1954) considering the amount of free amino acids liberated from the protein substances as a measure of proteolytic activity.

4% w/v homogenates were prepared in cold distilled water. The homogenates were centrifuged at 1000 rpm for 10 minutes. The supernatant was used as enzyme source. The reaction mixture in a volume of 2 ml contained 100 μ moles of phosphate buffer (pH 7.4), 20 mg of heat denatured haemoglobin as substrate and 0.5ml of the supernatant. The contents were incubated at 37°C for 30 minutes and the reaction was stopped by the addition of 2 ml of 10% TCA. Zero Time Controls were conducted by adding 2 ml of 10% TCA prior to the addition of enzyme source. The contents of the samples were filtered and the free amino acid level was determined in the filtrates. To 0.5 ml of aliquot of the filtrate, 2 ml of ninhydrin reagent was added. The contents were heated in boiling water bath for 5 minutes and cooled. The volume was made up to 10 ml with distilled water and read at 570 nm against a reagent blank in a spectrophotometer. All the samples were corrected with Zero time controls.

The proteolytic activity was expressed as μ moles of tyrosine equivalents / mg protein / hr.

3.4 Estimation of aspartate aminotransferase (AST)

The activity of aspartate aminotransferase (AST) was assayed by the colorimetric method of Reitman and Frankel (1957) as described by Bergmeyer and Bernt (1965).

33

2 % w/v tissue homogenates of the selected tissues were prepared in 0.25 M ice cold sucrose solution. The homogenates were centrifuged at 1000xg for 15 minutes and supernatant was used for the enzyme assay. The incubation mixture of 2.0 ml contained 100 μ moles of phosphate buffer (Na_2HPO_4 + NaH_2PO_4) (pH 7.4), 100μ moles of L-aspartatic acid, 2 μ moles of α-keto glutarate and 0.5 ml of supernatant as enzyme source. The test tubes were incubated for 30 minutes at 37°C and the reaction was stopped by the addition of 1 ml of ketone reagent (0.001 M, 2,4-dinitrophenyl hydrazine solution in 1 N HCl) and the contents were allowed to stay at room temperature for 20 minutes. After 20 minutes 10 ml of 0.4 N NaOH was added. The developed color was read at 545 nm in a spectrophotometer against a reagent blank.

The enzyme activity was expressed as μ moles of pyruvate formed / mg protein / hr.

3.5 Estimation of alanine aminotransferase (ALAT)

The activity of alanine aminotransferase (ALAT) was assayed by the colorimetric method of Reitman and Frankel (1957) as described by Bergmeyer and Bernt (1965).

The incubation mixture of 2 ml contained 100 μ moles of DL-alanine, 100 μ moles of phosphate buffer (pH 7.4), 2 μ moles of α-ketoglutarate and 0.5 ml of the supernatant of the homogenate 2% w/v prepared in 0.25 M ice-cold sucrose solution, as enzyme source. The reaction mixture was incubated at 37°C for 30 minutes. The reaction was stopped by the addition of 1.0 ml of 2, 4-dinitrophenyl hydrazine solution prepared in 1 N HCl (ketone reagent). The color was developed by the addition of NaOH. The optical density was measured at 545 nm in a spectrophotometer against a reagent blank. The enzyme activity was expressed as μ moles of pyruvate formed / mg protein / hr.

34

3.6 Estimation of Glutamate dehydrogenase (GDH)

The activity of GDH was assayed by the method of Lee and Lardy (1965).

3% w/v tissue homogenate was prepared in ice cold 0.25M sucrose solution and centrifuged at 1000xg for 15 minutes. The supernatant was used as enzyme source. The reaction mixture in a volume of 2ml contained 100 μ moles of phosphate buffer (pH 7.2), 4.0 μ moles of sodium glutamate, 0.1 μ moles of NAD, 4 μ moles of INT and 0.2 ml of enzyme source. The reaction mixture was incubated at 37°C for 30 minutes and the reaction was stopped by adding 5 ml of glacial acetic acid. Zero-Time Controls were maintained by adding 5 ml of glacial acetic acid prior to the addition of homogenate. The formazon formed was extracted overnight in 5 ml of cold toluene. The intensity of color developed was read at 495 nm against a reagent blank in a spectrophotometer.

The enzyme activity was expressed as μ moles of formazon formed / mg protein / hr.

3.7 Estimation of ammonia

Ammonia was estimated by the method of Bergmeyer (1965).

Homogenates of tissues were prepared in 10 % TCA medium and the homogenates were centrifuged at 1000xg for 15 minutes and to the clear supernatant 2 ml of 15% sodium hydroxide was added and to this 0.5 ml of Nessler's reagent was added and the optical density of the color was read immediately at 495 nm in a spectrophotometer against a reagent blank.

Ammonia content was expressed as μ moles of ammonia/g wet weight of tissue.

3.8 Estimation of urea

Urea was estimated by the diacetylmonoxime method as described by Natelson (1971).

35

Tissues were isolated and homogenates were prepared in 15% perchloric acid. To 0.5ml of the supernatant, 1 ml of acid mix (1:3 sulfuric acid: phosphoric acid) and 0.5ml of 2% diacetylmonoxime were added and vortexed. The contents were boiled in a water bath for 30 minutes and cooled to the laboratory temperature immediately. The optical density of the samples was read at 480 nm against the reagent blank in a spectrophotometer. Urea content was expressed as μ moles of urea/g wet weight of tissue.

3.B AGAROSE GEL ELECTROPHORESIS

Modified protocol of phenol-chloroform extraction

Agarose gel electrophoresis was estimated by modified protocol of phenol-chloroform extraction as described by Nelson (2008).

Tissues of rat samples were placed in microtubes, treated with 550 μL of lysis buffer solution (50 mM of Tris-HCl pH 8.0, 50 mM of EDTA, 100 mM of NaCl) plus 1% of SDS and 7 μL of 200 μg·mL-1 of proteinase K, and then were incubated in a thermo regulated bath at 50°C for 12h. Then, the DNA was purified with two separate extractions of phenol (250 μL) and three separated extractions with chloroform (250 μL). The DNA obtained was precipitated with 750 μL of absolute cold ethanol and with 300 μL of sodium acetate, and then it was incubated at -20°C for 2 h. The DNA samples were centrifuged with 700 μL of 70% ethanol, and resuspended in TE buffer (10 mM of Tris Hcl pH 8.0 and 1mM of EDTA before being treated with 30 μg·mL-1 of RNAse. Later, the DNA obtained was incubated in a water bath at 37°C for 40 min, and then kept at -20°C.

DNA FRAGMENTATION ASSAY

DNA fragmentation assay was estimated by diphenylamine method as described by Bahman Maroufi (2005).

The extracted DNA was transferred to a micro centrifuge tube. The DNA were lysed with 0.5ml ice cold lysis buffer (10mM Tris HCL, pH 7.5, containing 1 mM EDTA and 0.2% Triton X-100). Fragmented DNA was separated from intact chromatin by centrifugation for 10 min at 13000×g, 4°c (preparation B). The supernatant was carefully transferred to a test tube (preparation A). 0.5 ml of the lysis buffer was added to pellet containing preparation B. 0.5 ml of 25% trichloroacetic acid (TCA) was added to A and B preparations and vortex vigorously. The tubes were placed at 4°c and left the precipitate over night. The precipitates were centrifuged for 10 min at 13000×g. The supernatants were aspirated and discarded. 80 µl of 5% TCA was added to each pellet and the DNA was hydrolyzed by heating for 20 min at 83° c in a water bath. 160 µl of diphenylamine solution was added to the test tubes and to a blank containing 80 µl 5% TCA. All the test tubes were vortexed and then left overnight at room temperature. The collected supernatants were transferred to 96 well plate and optical densities were read at 620 nm by ELIZA reader. The percentage of fragmented DNA was calculated according to the following formula:

$$\% \text{ fragmented DNA} = \frac{OD\,620\,\text{tube A}}{OD\,620\,\text{tube A} + OD\,620\,\text{tube B}}$$

4. DETOXIFICATION ENZYMES

4.1. Estimation of Xanthine oxidase

Xanthine oxidase activity was estimated by the dye reduction method of Srikanthan and Krishnamoorthy (1955).

The enzyme source and 100 mM sodium phosphate buffer (pH 7.4) and 50 µ M of INT was added and the reaction was initiated by the addition of enzyme source and was incubated at 37°C for 30 minutes. The reaction was stopped by the addition of 5 ml of glacial acetic acid and the formazon formed overnight was extracted in toluene and the optical density was read at 495 nm against toluene blank. The activity was expressed as µM of formazon formed /mg protein / hour.

4.2. Estimation of Superoxide dismutase

Superoxide dismutase activity was determined according to the method of Beachamp and Fridovich (1971).

37

The activity of SOD was assayed by the reduction of nitro blue tetrazolium. Here, the superoxide was produced by riboflavin mediated photochemical reaction system. Different tissues were homogenized in ice cold 50 mM phosphate buffer (pH 7.0) containing 0.1 mM EDTA to give 5% homogenate (w/v) and the homogenate were centrifuged at 10,000 rpm for 10 minutes at 0 °C in cold centrifuge and the supernatant was separated and used for enzyme assay. The reaction mixture contained 1.7 ml of phosphate buffer (pH 7.8) and 150 ml EDTA (10 mM) and 600 ml methionine (130 mM) and 300 ml nitro blue tetrazolium (750 mM) and the enzyme source. The reaction was initiated by the addition of riboflavin and the samples were placed under 15 watts fluorescence bulb for 30 minutes and the absorbance was taken at 560 nm against reagent blank kept in a dark place. A system, devoid of any superoxide radical scavenger was used as a positive control to compare the results.

The activity of the enzyme was expressed as units/mg protein.

4.3. Estimation of Catalase activity

Catalase activity was measured by a slightly modified version of Aebi (1984) at room temperature.

Different tissues were homogenized in ice-cold 50 mM phosphate buffer (pH 7.0) containing 0.1 mM EDTA to give 5% homogenate (w/v). The homogenates were centrifuged at 10,000 rpm for 10 minutes at 0 °C in cold centrifuge. The resulting supernatant was used as an enzyme source. 10 μl of 100% ethyl alcohol was added to 100 μl tissue extract and then placed in an ice bath for 30 min. After 30 minutes, the tubes were kept at room temperature followed by the addition of 100 μl of Triton X- 100 RS and the cuvette was taken and add 200 μl of phosphate buffer and 50μl of tissue extract and 250 μl of 0.066 M H_2O_2 (in phosphate buffer) was added. The optical density was measured at 240 nm in a UV spectrophotometer. The molar extinction coefficient of 43.6 μc.m[-1] was used to determine catalase activity. One unit of activity is equal to the moles of H_2O_2 degraded / mg protein / min.

5. HISTOLOGY

Histological examinations of the tissues were followed according to Humason (1972).

Light microscopy

Tissues were isolated from control and snake *Naja naja* venom treated rats. They were gently rinsed with a physiological saline solution (0.9% NaCl) to remove blood and debris adhering to the tissues. They were fixed in 5% formalin for 30 minutes. The fixative was removed by washing through running tap water overnight. After dehydrating through a graded series of alcohols, the tissues were cleared in methyl benzoate, embedded in paraffin wax. Sections were cut at 6μ thickness and stained with hematoxylin (Harris, 1900) and counter stained with eosin (dissolved in 95% alcohol). After dehydration and clearing the sections were mounted with DPX and observed under the microscope.

VALIDITY OF EXPERIMENTAL PROCEDURES

Procurement of chemicals

1. All the chemicals used in this study were of analog grade and were procured from the following companies. i) Sigma, ii) BDH, iii) E. Merch, iv) Loba, v) Merck, vi) Kochlite

2. **Aliquots for assay:** Aliquots were selected such that initial rates were approximated as nearly as possible yet providing sufficient product to fall in a convenient range for Spectrophotometric measurement.

3. **Enzyme units:** Enzyme activities were expressed in standard units i.e., μ moles of product formed or substrate utilized per mg protein per minute.

4. **Substrate requirements:** All the enzyme assays were made under the conditions following zero order kinetics.

5. **Lambert-Beer's law:** Almost all the products of the reactions were

39

measured by colorimetric procedures in which the optical density (absorbance) of the resulting colored complexes was proportional to the concentration of the reaction products.

6. **Enzyme nomenclature:** The nomenclature of the enzyme used in the present context was according to the report of the commission on enzyme of the International Union of Biochemistry.

7. **Assay of dehydrogenases by using INT:** The advantage of using tetra-zolium salts as electron acceptors are :

    1.   Tetrazolium salts give a stable color on reduction.

    2.   They are highly soluble in aqueous solution.

    3.   They can be reduced both aerobically and anaerobically.

    4.   They have high redox potential which makes the reduction easier.

    5.   They are freely permeable through membrane.

Various tetrazolium salts receive electrons from various sites of electron transport system (Nachlas *et al.*, 1960). This is due to the inherent difference in the redox potentials of tetrazolium salts. The introduction of P-Nitro-Phenyl group in N_2 phenyl region was observed to increase the efficiency of the dye by increasing its redox potential. Karmarker *et al.*, (1959) reported INT was superior to most of tetrazolium salt as an electron acceptor for the assay of dehydrogenase.

8. **Statistical treatment of the data**

The mean, standard deviation (SD), percent change and one way analysis of variance (ANOVA) (Steel and Torrle, 1960) were performed using the SPSS package programming techniques on "Intel Core 2 Duo Processor" personnel computer. Probability values less than 0.05 were considered significant (Snedecor and Cochran, 1968).

Chapter-I

Toxicity Evaluation

41

Toxicology, although an old branch of science, is a branch of medical science that deals with the nature, effects and properties and the detection of poisons. It is defined as any harmful effect of chemical or a drug on a test organism. Extremely high concentrations of these chemicals are more toxic to the biological systems. Evaluation of toxicity of a chemical therefore is necessary to know, because it would help us to know its potentiality so that it could be possible to desire more powerful formulations. Unconscious and reckless handling of chemicals resulted in several disastrous incidences of pollution and accidental poisoning. Hence man has recognized the need for better control of the present use and future development of chemicals. In recent years it has become a normal practice to test all new chemicals for the toxicity before they could reach in consumers. So, evaluation of the toxicity of the chemical is an important tool that helps in the field of toxicology.

The term toxicity and hazard have been defined as a determinable value which is the capacity of a substance to produce injury and an indeterminable value which is the probability that injury will result from the use of substance in the proposed quantity and manner (Cartson, 1962).

Toxicity means the intrinsic capacity of a chemical substance or a mixture of substances to induce injury. Hazard means the observed toxic manifestation(s) induced by a known quantity of a substance under known exposure conditions. The term is frequently used interchangeably with "intrinsic toxicity".

A major purpose of the toxicological investigations to provide a basis for estimating the maximum dose that may be tolerated by animals through out their life time without manifesting any adverse effects (Gralla, 1981). The choice of the animal model to be used in toxicity studies appears critical as different tissues may be the target of the drug's toxic effect in different animal species (Roncaglioni *et al.*, 1982).

42

LD_{50} is the amount of a material, given all at once, which causes the death of 50% (one half) of a group of test animals. The LD_{50} is one way to measure the short-term poisoning potential (acute toxicity) of a material. Toxicologists can use many kinds of animals but most often testing is done with rats and mice. It is usually expressed as the amount of chemical administered (e.g., milligrams) per 100 grams (for smaller animals) or per kilogram (for bigger test subjects) of the body weight of the test animal. The LD_{50} can be found for any route of entry or administration but dermal (applied to the skin) and oral (given by mouth) administration methods are the most common.

Inhalation and skin absorption are the most common routes by which workplace chemicals enter the body. Thus, the most relevant from the occupational exposure viewpoint are the inhalation and skin application tests. Despite this fact, the most frequently performed lethality study is the oral LD_{50}. This difference occurs because giving chemicals to animals by mouth is much easier and less expensive than other techniques. However, the results of oral studies are important for drugs, food poisonings, and accidental domestic poisonings. Oral occupational poisonings might occur by contamination of food or cigarettes from unwashed hands, and by accidental swallowing.

Snake venom and other toxins strength is measured using the LD_{50} (lethal dose 50%) test. It involves dosing several groups of animals with a substance either by mouth (force-feeding), by injection, via the skin or by inhalation. Animals used are mice, rats, rabbits, cats, dogs, monkeys, fish and birds etc. The finishing point of the test is when half of the animals in the group have died. The amount of test substance that kills half the animals gives the LD_{50} figure. LD_{50} figures are used, in theory, to indicate the standard toxicity value for each chemical.

The animals may be given the chemical compound to be tested through any of the major portal routes i.e., dermal, oral and respiratory. Studies also include skin and eye irritation experiments. However the chances of poisoning are more through food and water in animals particularly in human beings. Considerable number of accidental poisonings and suicide attempts happened when the pesticides (or) toxic chemicals were taken orally. Moreover, oral method is considered as the best to evaluate the toxicity in vertebrates. In evaluating the oral toxicity, consideration must be given to all aspects of biochemical, physiological, pharmacological and morphological changes.

This test has many downfalls and has been called "crude and unscientific". LD_{50} figures for a single chemical can vary enormously according to the species, strain, age, gender and even diet of the animals used in the test.

Venoms of all snake specimens were pooled and dissolved in saline (final concentration: 10 mg/ml). All the venom samples were stored at 4 °C to avoid any disruption of their natural toxic properties.

In general, if the immediate toxicity is similar in all of the different animals tested, the degree of immediate toxicity will probably be similar for humans. When the LD_{50} values are different for various animal species, one has to make approximations and assumptions when estimating the probable lethal dose for man. Different LD_{50} values for different snake venoms are given in table 1.2-1.3. Special calculations are used when translating animal LD_{50} values to possible lethal dose values for humans.

In general the rat and mouse are normally used for oral toxicity test, because they have digestive system geared to the same sort of diet as man in many respects, and has similar metabolic responses. Generally the lethality of a venom to a particular animal species is expressed in terms of mortality and time. In case of terrestrial animals it is expressed in terms of lethal dose (LD) as mg/kg weight of animals. In aquatic animals the lethality is expressed in terms of lethal concentration (LC) as mg/litre (or) parts per million (ppm) (or) parts per billion (ppb) while conducting the toxicological studies with the administration of toxicants.

In the present study the toxicity of snake *Naja naja* venom in albino rats was determined, to ascertain the LD_{50} and sub lethal doses. Determinations of these doses are essential to fix the dosing schedule for the development of tolerance and to examine the consequent changes in biochemical parameters and related histopathology. The signs and symptoms were also monitored throughout the period of dosing schedule.

Dose-response relationship

The dosage of any compound is always a decisive factor in determining its effects. Hence it is important to measure the toxicity i.e., the determination of the dose or concentration at which toxicant produces harmful response on a target organism. Dose refers to a stated quantity or concentration of a substance to which an organism is exposed. It is most commonly expressed as the amount of test substance per unit weight of test animal (e.g. mg/kg body weight) (Oecd, 2000). Dosages often involve the dimension of time (e.g. mg/kg/day).

Dose-response relationship means the correlative association existing between the dose administered and the response (effect) or spectrum of responses that is obtained. The concept expressed by this term is indispensable to the identification, evaluation, and interpretation of most pharmacological and toxicological responses to chemicals. The basic assumptions which underlie and support the concept are: (a) the observed response is a function of the concentration at a site; (b) the concentration at a site is a function of the dose, and (c) response and dose are causally related (Eaton and Klaassen, 1996). The existence of a dose-response relationship for a particular biological or toxicological response (effect) provide a defensible conclusion that the pressure is a result of exposure to a known substance.

The venom which reach the experimental animals must be provided to assess the risk to humans and animals if they are exposed to acutely high concentrations and the knowledge of venom intoxification is predominantly based on LD_{50} determination involving smaller number of animals. This parameter is used by the majority of researchers to characterize the toxicity of substances.

Results

The data was computed according to Probit analysis (Finney, 1971), and LD_{50} values were determined. The animals were exposed to different concent-rations of snake *Naja naja* venom, showed no mortality up to 0.30 mg/kg body weight, 16.67% mortality at 0.32 mg/kg body weight, 25% mortality at 0.34 mg/kg body weight, 33.33% mortality at 0.36 mg/kg body weight, 50% mortality at 0.38 mg/kg body weight, 67% mortality at 0.40 mg/kg body weight, 83% mortality at 0.42 mg/kg body weight, 100% mortality at 0.44 mg/kg body weight and 0.46 mg/kg body weight observed (Table-1.1).

The computation of percent mortality against different log concentrations of the snake *Naja naja* venom is given a typical sigmoid curve (Fig. 1.1). The LD_{50} value obtained from the sigmoid curve is 0.38 mg/kg body weight for 48 hours. The probit mortality of the albino rats were calculated from percent mortality, when the probit mortality was plotted against log concentrations of the snake *Naja naja* venom a straight line was obtained (Fig. 1.2). The LD_{50} value obtained from this straight line graph is 0.38 mg/kg body weight. After the administration of snake *Naja naja* venom, the rats lost their normal activity and showed different abnormalities when compared to control animals. Since the snake *Naja naja* venom is highly toxic, $1/50^{th}$ of LD_{50} i.e., 0.768 mg/kg body weight was taken as sub lethal dose for experimentation.

Discussion

Snake *Naja naja* venom is a highly poisonous venom used to kill different types of animals including human beings. The LD_{50} values of snake *Naja naja* species are summarized in Table 1.2. Covacevich and Wombey (1976) showed that the small scaled snake (*Parademansia microlepidotus*) was taxonomically distinct and highly venomous snake.

Broad *et al.*, (1979) demonstrated two-directional polyacrylamide gel electrophoresis, that venom from the two snakes had characteristically distinct electrophoretograms and this has been used as a tool to distinguish the species of different snakes in taxonomy. Toxicity studies with venom from *P. microlepidotus* showed that it was considerably more toxic than any other Australian snake venom. It was decided therefore to determine the exact toxicity status of *P. microlepidotus* venom related to other toxic terrestrial snake venoms. Broad *et al.*, (1979) reported different snake LD_{50} values in mice with saline as medium and 0.1% bovine serum albumin in saline (Table 1.3).

Researchers have usually carried out LD_{50} calculations according to Reed and Muench (1958), Litchfield and Wilcoxon (1949) method. Nevertheless, the information on the lethality of snake venoms is required although in many cases it may result in unnecessary waste of experimental animals (Zbinden and Flurry, 1981, Brown, 1985). The LD_{50} test introduced by Trevan (1927) has gained acceptance as a measure of acute toxicity.

Meier and Theakston (1986) calculated the approximate LD_{50} values of some snake venoms by using the method of Baccari (1949) as modified by Molinengo (1979). Several researchers have studied the biochemical and pharmacological effects of venoms from different species of snakes from different localities (Al-Asmari, 2005). Meier and Theakston (1986) applied this technique and got consistent results. Alam and Ali (1998) determined LD_{50} of cobra venom. Little information is available on the pharmacological effects of venom of *Echis carinatus* (Saw-scaled viper) (Zahra *et al.*, 2005). The LD_{50} of each snake species was determined according to the mathematical scheme adopted by Meier and Theakston (1986).

The toxicity evaluation of Snake *Naja naja* venom is 38.4 mg/kg. Since the venom is highly poisonous, $1/50^{th}$ of the LD_{50} (0.768 mg/kg) was selected as sub lethal dose in the present study.

Table: 1.1   Mortality of Albino rats administered with different doses of Snake *Naja naja* venom.

Sl. No.	Dose mg/Kg	Log Concentration	Number of animals exposed	Number of animals died	Percent Kill	Probit Kill
1.	0.32	0.32109	12	2	16.67	4.05
2.	0.34	0.34109	12	3	25.00	4.33
3.	0.36	0.36109	12	4	33.33	4.56
4.	0.38	0.38109	12	6	50.00	5.00
5.	0.40	0.40109	12	8	67.00	5.44
6.	0.42	0.42109	12	10	83.00	5.95
7.	0.44	0.44109	12	12	100.00	8.09
8.	0.46	0.46109	12	12	100.00	8.09

Fig.1.1:   **Sigmoid "Graded Response" curve showing the relation between the log concentrations of Snake *Naja naja* Venom and percent mortality of the albino rats.**

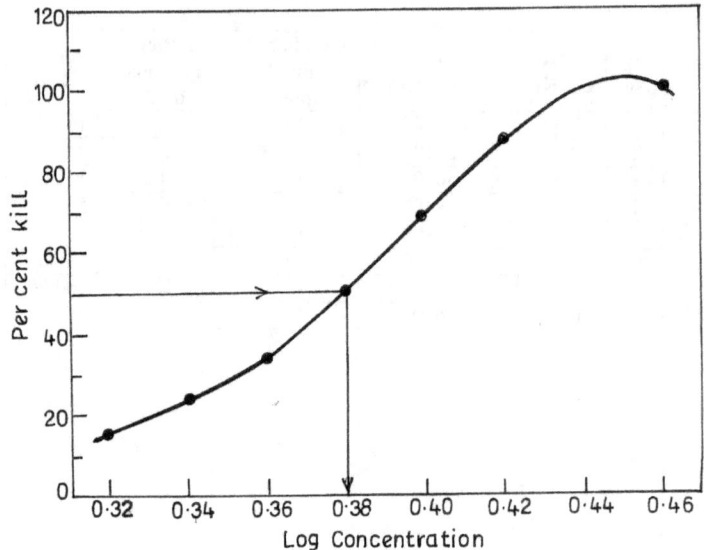

Fig.1.2 :   Graph Showing straight line relation between the log concentration of Snake *Naja naja* venom and probit mortality of the albino rat.

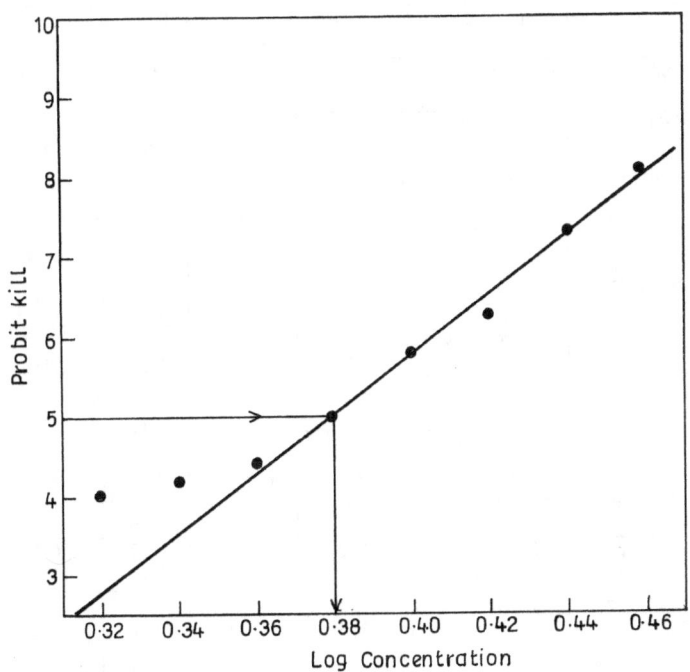

Table: 1.2 LD_{50} Scores for various snakes

Species	Common name	Sub-cutaneous mg/kg	Intra-muscular mg/kg	Intra-venous mg/kg	Intra - peritoneal mg/kg	Venom Yield mg
Naja atra	Chinese cobra	0.29		0.345		
Naja haje	Egyptian cobra	1.15		0.96	0.185	175.0-300.0
Naja kaouthia	Monocled cobra			0.373	0.225	
Naja melanoleuca	Forest cobra			0.6	0.324	500
Naja naja	Spectacled cobra	0.45		0.35	0.315	
Naja nigricollis	Black necked spitting cobra		0.44	1.15	0.4	200.0-350.0
Naja nivea	Cape cobra	0.72		0.57	0.4	
Naja oxiana	Central Asian cobra			0.96		
Naja pallida	Red spitting cobra				2	

Table : 1.3   **Snake venom LD$_{50}$ determinations in mice (18-21 g) by subcutaneous injection**

Snake (common name in parenthesis)	LD$_{50}$ mg/kg (95% confidence (limits))			
	Saline		0·1% bovine serum albumin in saline	
Parademansia microlepidotus (Small-scaled snake)	0·025	(0·020–0·029)	0·010	(0·007–0·014)
Pseudonaja textilis (Brown snake)	0·053	(0·041–0·065)	0·041	(0·033–0·051)
Oxyuranus scutellatus (Taipan)	0·099	(0·081–0·123)	0·064	(0·052–0·078)
Notechis scutatus (Tiger snake)	0·118	(0·095–0·146)	0·118	(0·088–0·157)
Notechis ater niger (Reevesby Island Tiger snake)	0·131	(0·107–0·163)	0·099	(0·083–0·120)
Enhydrina schistosa (Beaked sea snake)	0·164	(0·149–0·185)	0·173	(0·144–0·210)
Notechis ater occidentalis (Western Australian Tiger snake)	0·194	(0·161–0·234)	0·124	(0·102–0·152)
Notechis ater serventyi (Chappell Island Tiger snake)	0·338	(0·278–0·414)	0·271	(0·220–0·335)
Acanthophis antarcticus (Death adder)	0·400	(0·336–0·472)	0·338	(0·278–0·413)
Pseudonaja nuchalis (Gwardar)	0·473	(0·393–0·570)	0·338	(0·271–0·423)
Austrelaps superba, Australia (Australian Copperhead)	0·560	(0·448–0·700)	0·500	(0·415–0·605)
Naja naja (Indian Cobra)	0·565	(0·450–0·705)	0·500	(0·415–0·605)
Pseudonaja affinis (Dugite)	0·660	(0·550–0·800)	0·560	(0·505–0·620)
Pseudechis papuanus (Papuan black snake)	1·09	(0·865–1·35)	1·36	(1·23–1·51)
Hoplocephalus stephensii (Yellow banded snake)	1·36	(1·12–1·66)	1·44	(1·27–1·63)
Tropidechis carinatus (Rough scaled snake)	1·36	(1·19–1·56)	1·09	(0·980–1·21)
Ophiophagus hannah (King Cobra)	1·80	(1·50–2·18)	1·91	(1·55–2·37)
Pseudechis guttatus (Blue-Bellied black snake)	2·13	(1·79–2·51)	1·53	(1·24–1·89)
Pseudechis colletti (Collett's snake)	2·38	(2·08–2·74)	Not done	
Pseudechis australis (King brown snake)	2·38	(1·93–2·92)	1·91	(1·57–2·33)
Pseudechis porphyriacus (Red-Bellied black snake)	2·52	(2·09–3·04)	2.53	(2·05–3·14)
Cryptophis nigrescens (Small-Eyed snake)	2·67	(2·40–2·96)	Not done	
Crotalus adamanteus (Eastern Diamond-Back Rattlesnake)	11·4	(9·10–14·25)	7·70	(6·30–9·40)
Demansia olivacea (Spotted snake)	>14·2		Not done	
Bothrops atrox (Barba amarilla)	>27·8		Not done	

Chapter-II

Haematology

Blood is a specialized bodily fluid that delivers necessary substances to the body's cells such as nutrients, oxygen and transports of waste products away from those of same cells. The cells of the tissue of the body are in contact with body fluids which in turn are in equilibrium with the fluid portion of the blood. Blood is the most important body fluid that governs vital functions of the body like respiration, circulation, excretion, osmotic balance and the transport of metabolic substance. Circulation of the blood within the cardiovascular system is essential for transportation of gases, nutrients, minerals, metabolic products and hormones between different organs.

Haematology, the study of blood, is a basic medicinal science. In this discipline, the fundamental concepts of biology and chemistry are applied to the medical diagnosis in treatment of various disorders of disease related to the blood. Haematological indices vary from animal to animal and in some animals at different stages of life. At birth the haemoglobin content is higher during any other stage of life. The erythrocyte count and hemoglobin concentration gradually rise in the adult levels by the time of puberty with a characteristic low level in females than in males (Hawkins *et al.,* 1954).

Some of the poisonous snakes are Cobra (*Naja naja*), Russell's Viper (*Vipera russelli*), Saw-Scaled Viper (*Echis carinatus*) and Kraits. The deaths due to bites by these snakes amount to 10,000 to 12,000 annually in the subcontinent of India and Pakistan. Venom of these snakes contain many active components which act upon the blood constituents, nervous system and other systems of the body. Cobra venom is characterized predominantly by its neurotoxic effect. This neurotoxic effect is due to toxin, a protein present in the cobra venom. The haematological abnormalities produced by cobra bite are variable. In about 50% cases of cobra bite, moderate neutrophilic effect has been reported. No direct action of cobra venom on coagulation have been observed. Cobra venom has got mild accelerating effect on the lysis of normal euglobulin clot, indicating a fibrinolytic action. However one of the cobras i.e., *Naja nigricollis* has got antifibrinolyitc activity.

The effect of snake *Naja naja* venom on haematological parameters were extensively studied by different authors. The snake venom viper and cobra venom on blood (Pradeep kumar and Basheer, 2011), *Cerastes vipera* crude venom on plasma and tissue metabolites (Ibrahim Al-Jammaz, 2002), *Walterinnesia aegyptia* venom on, serum and tissue metabolites and some enzymes (Ibrahim Al-Jammaz *et al.,* 1994), cobra venom, on blood coagulation, platelets and fibrinolysis (Rehman *et al.,* 2006), *Cerastes cerastes gasperettii* venom on, blood cells count (Al-Sadoon and haffor, 2005) were studied. Another commonest serious effect of cobra poisoning in man is tissue necrosis.

The biochemical changes after envenomation being of vital importance, the present study aims to investigate the biochemical parameters such as RBC, Hb, MCV, MCH, MCHC, PCV, WBC, neutrophils, lymphocytes, monocytes, eosinophils, basophils, serum creatinine, cholestrol, triglycerides, serum total proteins, alkaline phosphatase, gamma glutamyl transferase, serum sodium, serum potassium, serum calcium, serum phosphorous, serum amylase, lactate dehydrogenase levels in Albino rat, resulting in the envenomation of snake *Naja naja* venom with in three successive days of 24 hrs, 48 hrs, 72 hrs respectively.

RESULTS

HAEMOGRAM

In the present investigation the effect of snake *Naja naja* venom, on the blood is determined in albino rats. The changes in haematology of albino rats as a result of snake *Naja naja* envenomation in 24 hrs, 48 hrs, 72 hrs are indicated in (Table 2.1 to 2.5 and Fig. 2.1 to 2.5).

Oral administration of snake *Naja naja* venom produced statistically significant (P<0.01) decrease in RBC, HB, PCV, MCV, MCH, MCHC, Platelet count(PC) but WBC shows increased level in 24 hrs, 48 hrs, 72 hrs of venom injection respectively (Table 2.1 and Fig 2.1). The red cell indicators like mean corpuscular volume (MCV), mean corpuscular hemoglobin (MCH) and mean corpuscular hemoglobin concentration (MCHC) are dependent on the RBC count, HB concentration and PCV values. MCV, MCH and MCHC and PC showed statistically significant (P<0.01) decrease in 24 hrs and 48 hrs and 72 hrs of intoxification in albino rats. WBC count shows statistically significant (P<0.01) increase in animals of experimental group compared to the control group.

HAEMOGRAM OF DIFFERENTIAL COUNT

The differential count (Table 2.2 and Fig.2.2) regarding lymphocytes, monocytes, eosinophils, basophils showed statistically significant (P<0.01) increase in all doses of animals compared to the control group. The neutrophils showed a statistically significant (P<0.01) decrease in all experimental animals. The slightly elevated eosinophils in animals of experimental group differ statistically from that observed in the control group. In the case of basophils no change is observed in all groups of experimental animals.

HAEMOGRAM OF CREATININE AND LIPID PROFILE

Serum creatinine levels (Table 2.3 and Fig.2.3) in the blood were increased from control to 24hrs, 48hrs, 72hrs of envenomated rats. The cholesterol and triglycerides in the serum of blood were reduced from control to 24 hrs, 48 hrs, 72 hrs in *Naja naja* envenomated rats.

HAEMOGRAM OF LIVER FUNCTION TEST (LFT)

The liver function test (LFT) (Table 2.4 and Fig.2.4) was done in the blood. Alkaline phosphatase and Gamma glutamyl transpeptidase (GGTP) were tested and there was a significant increase from control to 24 hrs, 48 hrs, 72 hrs, but where as the serum total proteins (STP) were significantly decreased from control to 24 hrs, 48 hrs, 72 hrs of envenomated rats (Table 2.4 and Fig.2.4).

HAEMOGRAM OF ELECTROLYTES

Serum electrolytes in the blood were investigated after envenomation of snake *Naja naja* venom, the result indicates that levels of sodium, potassium, calcium, serum amylase were increased significantly from control to 24hrs, 48hrs, 72hrs respectively, however serum phosphorous and lactate dehydrogenase activity were decreased from 24hrs, 48hrs, 72hrs in Albino rats after snake *Naja naja* envenomation (Table 2.5 and Fig. 2.5).

DISCUSSION

Junaid mohamood alam and Rashida qasim, (1993) worked on different snake venoms namely *Physalia* venom (vpp), *H.spiralis* venom (VHS), *H. cyanocinctus* venom (VHC), *H. lapermoides* venom (VCC), species and serum enzyme levels namely lactate dehydrogenase (LDH), gamma glutaryl transferase (GGT), alkaline phosphatase (ALP), amylase (AMY) have been estimated and it was found that these serum enzyme levels were increased significantly. The elevated levels of the serum enzymes such as Lactate dehydrogenase (LDH), γ- glutamyl transferase, alkaline phosphatase, amylase (AMY) clearly indicate the presence of necrosis or cellular damage in liver, kidney, brain, heart induced by the snake *Naja naja* venom. The elevated levels of serum enzymes of snake *Naja naja* venom confirmed about the severe morphological alterations in liver, kidney, brain, heart. Shiomi (1990) reported raised blood creatinine and blood urea levels in victims of sea snake bites with severe renal damage.

Junaid mohamood alam and Rashida qasim (1993) worked on different serum chemical components. Decreased levels of total lipids, triglycerides, cholesterol, total protein, albumin, and elevated levels of glucose, creatinine, total and direct bilirubin and blood urea nitrogen indicates the possibility of hepatitis, glomerulonephritis and renal failure, malabsorption syndrome and myocardial necrosis were observed after experimental envenomation by venomous marine animals.

Pradeep kumar and Basheer (2011) studied the blood samples of viper bite and cobra bite patients and analyzed the blood sample from each case taken as soon as bite was happened, further blood samples were collected at different intervals. The blood clotting time, blood urea, serum creatinine, serum sodium, serum potassium, serum calcium were estimated and it was observed that there was significant increase in above parameters. Results of calcium estimation showed that there was no significant increase in calcium levels in 24hrs, 48hrs, 72hrs of *Naja naja* envenomated rats. This is also observed in viper bite cases. This may be due to lowering of serum calcium concentration and calcium is not a valuable indicator for accessing the severity of snake bite cases. Results of potassium estimation showed that there was no significant increase in potassium levels in 24hrs, 48hrs, 72hrs of snake *Naja naja* envenomated rats. Similar cases were observed in viper bites. The results of sodium estimation showed that there was no significant increase in Na^+ levels in both cobra and viper bite (Pradeep kumar and Basheer, 2011). This increase may be due to the secondary effect of the renal failure.

Ibrahim and Al-jammaz (2003) studied the physiological effects of snake *Echis coloratus* crude venom in rat at different time intervals. The serum cholesterol was reduced gradually and serum triglycerides and alkaline phosphatase (ALP) were significantly increased. The variations in serum cholesterol was observed after 24hrs, 48hrs, 72hrs in envenomated rats. The results suggest that these variations could be time dependent (Ibrahim, Al-Jammaz, 2003). The total lipid contents of the tissues measured plasma cholesterol and triglyceride levels were significantly decreased in envenomated rats. This results suggests that the snake venom might have mobilized lipids from adipose tissues and other tissues, lipolytic enzymes which are present in many snake venoms, could have split tissue lipids with the liberation of free fatty acids. The disturbance in protein synthesis is due to cellular damage in hepatocytes together with haemorrhages in vital organs (Ibrahim and Al Jammaz, 2003). Moreover haemorrhages in vital organs together with increased vascular permeability, renal damage would further aggrevate the accompanying hypoprotenimia and hypoalbuminae (Ibrahim and Al Jammaz, 2003). The severe hepatocellular injuries, necrosis of hepatocytes and kidney have been suggested to be the result of the significant rise in alkaline phosphatase (ALP) levels after the envenomation of the snake *Naja naja* venom. The increased levels of serum alkaline phosphatase in *Naja naja* envenomation might be attributed to the destruction of the liver cells (Abdel-Nabi *et al.,*1993). The elevated activity of Alanine amino transferase (ALAT), Amylase together with the elevation of Lactate dehydrogenase (LDH) indicate the damage of liver, kidney, brain and heart. These findings are in agreement with previous reports of venom of other snake species.

Ibrahim and Al Jammaz *et al.,* (1994) studied the effects of *Walterinnesia aegyptia* venom on serum and tissue metabolites and on some enzyme activities in Albino rats. This venom caused a decrease in serum lipids, triacyl glycerols and total cholesterol with a decline in lactate dehydrogenase (LDH) activities. However, the inhibition of the lipolytic action of the venom may develop as a result of the possible presence of an enzyme inhibitor in the venom (Middleton *et al.*, 1964) which seems to be dominant when large doses of the venom have been applied. This might be indicated from the present observation of decreased serum lipids content, as well as in the liver, kidney, brain, heart, triacylglycerol, and total cholesterol contents. This could be supported by the observation of Braganca *et al.*, (1970) of a polypeptide inhibitor specific for phospholipase A in cobra venom. The decline in lactate dehydrogenase activity could well be due to the action of non-specific proteolytic enzymes present in the venom (Braganca and Quastel, 1953) but not due to the specific inhibitors present in the venom (Al-Jammaz, 1993).

Al-Jammaz *et al.,* (1993) reported the inhibited Lactate dehydrogenase enzyme (LDH) level in the liver and kidney of envenomated rats could be further evidence for a disturbance in pyruvate oxidation, as well as in the Kreb's cycle taking into consideration that these organs possess these main metabolic pathways. The elevated Lactate dehydrogenase (LDH) enzyme level in the brain and heart might suggest the prevalence of anaerobic conditions in such vital organs as a result of enhancing the metabolic cycle to restore energy loss following *Walterinnesia aegyptia* venom, as has been observed in case of the venoms of both *D. Polylipis* and *N. haje* (Venkateswarulu *et al.*, 1978) which have been reported to induce the same effects on enzyme activities, as those observed in the present study.

Variations in serum physiological parameters can be used as biomarkers for monitoring the functions of vital organs of envenomated victims. The reduction in total serum proteins and albumin and the rise in total serum bilirubin in envenomated rats are in accordance with observations of other investigators in this field (Fahim, *et al.*, 1998). The observed effects upon those parameters might suggest that the snake venom could have disturbed protein synthesis in hepatocytes due to cellular damage together with haemorrhges in vital organs leading to protein loss. Acute renal damage together with glomerular, tubular, interstitial and vascular lesions have been reported following various snake bite (Sitprija *et al.*, 1982). Moreover, haemorrhages in vital organs, together with increased vascular permeability were observed in the majority of viper and pit viper envenomation (Meier *et al.*, 1991). Such increased vascular permeability, together with renal damage would further aggravate the accompanying hypoproteinemia and hypoalbuminaemia. Furthermore, the rise in serum urea and creatinine are associated with the reduction of serum uric acid level observed in the present study, supports the proposed impairment of renal function. Similar observations were reported following various viper envenomation of rats (Abdel-Nabi *et al.*, 1997).

Several studies were undertaken to determine the effect of the venom of some members of the viper family on metabolism and important blood parameters of animals (Fahim, 1998). Though the venom of most viper snakes induces hyperglycemia in experimental animals (Fahim, 1998), other venoms were reported to induce hypoglycemia (Ibrahim Al-Jammaz, 2002).

In the present study due to snake venom envenomation a decrease in RBC, Hb, PCV, MCV, MCH, MCHC, platelet count, serum creatinine and increase in WBC, were observed. In differential count also, a gradual decrease in neutrophils, serum total proteins and increase in lymphocytes and monocytes, eosinophils, basophils, cholesterol, triglycerides, alkaline phosphatase, gamma glutamyl transferase, and electrolytes such as sodium, potassium, calcium, phosphorous, amylase, lactate dehydrogenase were observed. The change in haemogram indicates that the snake *Naja naja* venom alters the biochemical pathways and causes cellular damage.

Table 2.1: **Haemogram of control and snake *Naja naja* venom treated albino rats.**

Parameters	Control	24 Hours	48 Hours	72 Hours
RBC (cu. mm)				
Mean	8.685	5.596	5.783	4.943
SD	±0.021	±0.020	±0.029	±0.024
PC		(-35.592)	(-33.441)	(-43.111)
Hb/(g/100ml)				
Mean	16.387	13.094	11.996	9.695
SD	±0.016	±0.014	±0.020	±0.019
PC		(-20.091)	(-26.791)	(-40.833)
PCV(Percent)				
Mean	44.211	40.994	39.095	34.793
SD	±0.023	±0.018	±0.021	±0.024
PC		(-7.277)	(-11.572)	(-21.303)
MCV(fl)				
Mean	74.211	70.994	69.095	64.793
SD	±0.023	±0.018	±0.025	±0.024
PC		(-1.840)	(-2.531)	(-3.495)
MCH(pg)				
Mean	22.596	21.395	19.595	18.686
SD	±0.011	±0.018	±0.015	±0.019
PC		(-5.314)	(-13.281)	(-17.301)
MCHC(Percent)				
Mean	37.449	31.995	30.696	27.894
SD	±0.011	±0.017	±0.020	±0.023
PC		(-14.564)	(-18.032)	(-25.526)
Platelet count(Lakhs/cumm)				
Mean				
SD	3.344	2.835	1.815	1.496
PC	±0.021	±0.026	±0.016	±0.013
		(-15.235)	(-45.736)	(-55.280)
WBC(cu. mm)				
Mean	11156.630	12300.905	13451.203	14262.965
SD	±0.020	±0.019	±0.009	±0.021
PC		(10.256)	(20.567)	(27.843)

All the values are mean ± SD of six individual observations.
SD – Standard Deviation. PC – Percent change over control.

ONE WAY ANOVA

Source of Variation	Df	Mean of Significance							
		RBC	Hb	PCV	MCV	MCH	MCHC	PC	WBC
Between Groups	3	16.569**	46.493**	92.903**	92.903**	18.565**	96.513**	4.476**	110277 87.894**
Within Groups	20	0.001	0.000	0.000	0.000	0.000	0.000	0.000	0.000
Total	23								

NS: Not Significant,* Significant (P < 0.05), ** Highly Significant (P < 0.01)

Fig. 2.1 : Changes in haematological parameters in albinorats exposed to snake *Naja naja* venom

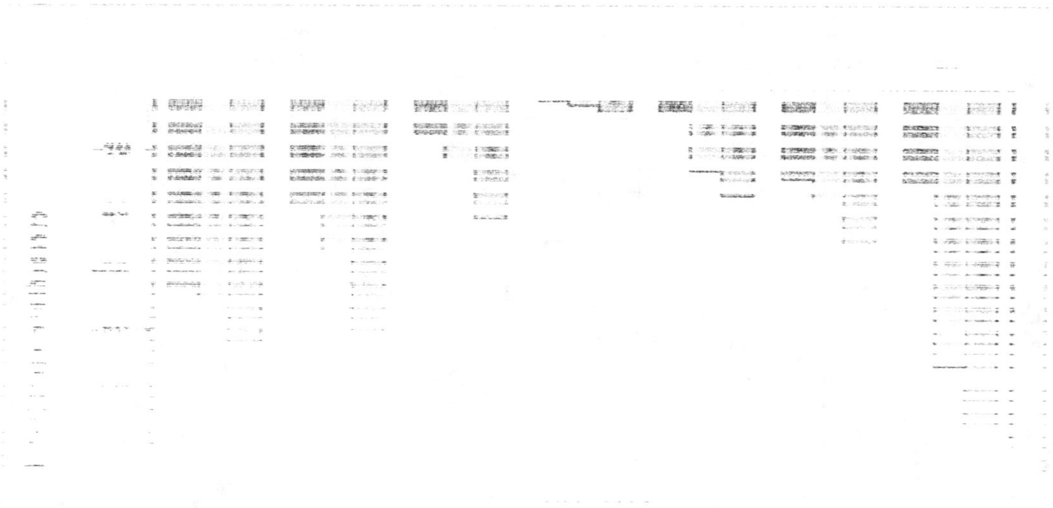

Table 2.2:   Haemogram of differential count of control and snake *Naja naja* venom treated albino rats.

Parameters	Control	24 Hours	48 Hours	72 Hours
Neutrophils				
Mean	20.108	18.686	16.387	14.996
SD	±0.022	±0.019	±0.016	±0.024
PC		(-7.071)	(-18.508)	(-25.423)
Lymphocytes				
Mean	474.992	751.992	1259.991	1399.995
SD	±0.024	±0.025	±0.027	±0.022
PC		(58.317)	(165.266)	(194.740)
Monocytes				
Mean	1.993	4.994	8.688	13.995
SD	±0.009	±0.024	±0.027	±0.015
PC		(148.060)	(335.926)	(602.191)
Eosinophils				
Mean	3.998	7.998	9.998	27.996
SD	±0.017	±0.018	±0.019	±0.014
PC		(100.021)	(150.050)	(600.183)
Basophils				
Mean	1.000	1.000	1.000	1.993
SD	±0.000	±0.000	±0.000	±0.009
PC		(0.000)	(0.000)	(99.300)

All the values are mean ± SD of six individual observations.
SD – Standard Deviation.
PC – Percent change over control.

ONE WAY ANOVA

Source of Variation	Df	Mean of Significance				
		Lymphocytes	Monocytes	Neutrophils	Eosinophils	Basophils
Between Groups	3	3369228.341[**]	482.501[**]	94.272[**]	2033.564[**]	4.478[**]
Within Groups	20	0.012	0.008	0.009	0.006	0.002
Total	23					

NS: Not Significant, * Significant (P < 0.05), ** Highly Significant (P <0.01)

Fig. 2.2 : Changes in haematological parameters in albino rats exposed to snake *Naja naja* venom

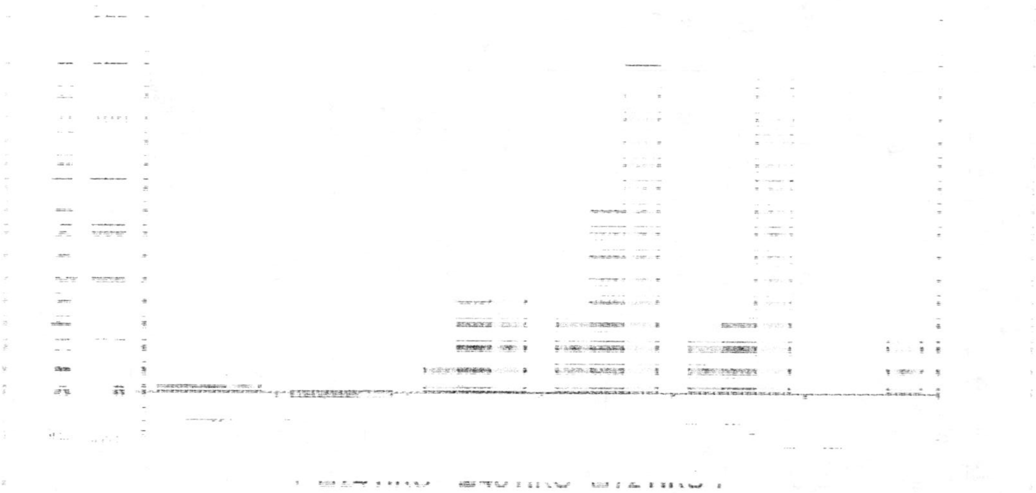

Table 2.3 :   Haemogram of Serum creatinine and lipid profile of control and snake *Naja naja* venom treated albino rats.

Parameters	Control	24 Hours	48 Hours	72 Hours
Sr.Creatinine(mg/dl)				
Mean	0.898	0.995	1.297	1.798
SD	±0.014	±0.021	±0.019	±0.018
PC		(10.824)	(44.486)	(100.316)
LIPID PROFILE				
Cholesterol(mg/dl)		59.997	58.996	56.997
Mean	63.997	±0.019	±0.012	±0.020
SD	±0.020	(-6.251)	(-7.814)	(-10.939)
PC				
Triglycerides(mg/dl)				
Mean	64.997	61.996	51.996	49.995
SD	±0.020	±0.017	±0.020	±0.022
PC		(-4.610)	(-20.002)	(-23.080)

All the values are mean ± SD of six individual observations.
SD – Standard Deviation.
PC – Percent change over control

ONE WAY ANOVA

Source of variation	Df	Mean of significance		
		Sr.creatinine	Cholestrol	Triglycerides
Between Groups	3	2.951[**]	156.042[**]	976.725[**]
Within Groups	20	0.006	0.006	0.008
Total	23			

NS:Not Significant, * Significant (P < 0.05), ** Highly Significant (P < 0.01)

Fig. 2.3 : Changes in serum creatinine and lipids of control and snake *Naja naja* venom treated albino rats.

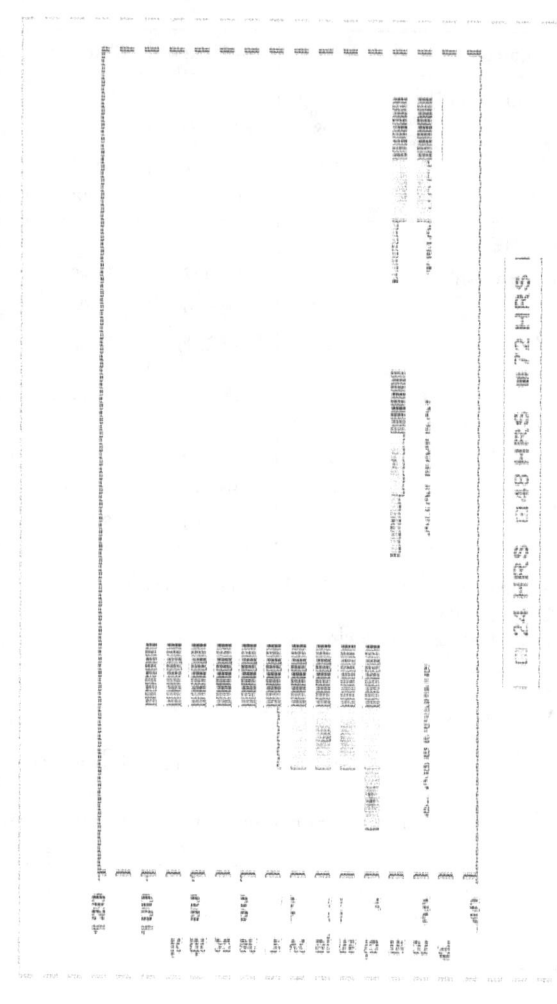

Table 2.4: Haemogram of Liver function tests (LFT) of control and snake *Naja naja* venom treated albino rats.

Parameters	Control	24 Hours	48 Hours	72 Hours
Sr.Total Proteins(gm/dl)				
Mean	8.103	7.808	7.704	7.205
SD	±0.009	±0.028	±0.015	±0.019
PC		(-3.641)	(-4.918)	(-11.077)
Alkaline phosphatase (ALP) **(IU/L)**				
Mean	905.003	915.010	963.995	972.008
SD	±0.009	±0.037	±0.019	±0.028
PC		(1.106)	(6.518)	(7.404)
Gamma glutamyl transferase (GGT) (IU/L)				
Mean	4.008	8.003	10.010	13.005
SD	±0.028	±0.009	±0.037	±0.019
PC		(99.668)	(149.782)	(224.517)

All the values are mean ± SD of six individual observations.
SD – Standard Deviation.
PC – Percent change over control.

ONE WAY ANOVA

Source of Variation	Df	Mean of Significance		
		Sr. Total Proteins	Alkaline phosphatase (ALP)	Gamma glutamyl transferase (G.G/T)
Between Groups	3	2.511**	20674.891**	256.455**
Within Groups	20	0.007	0.014	0.013
Total	23			

NS: Not Significant, *Significant ($P < 0.05$), ** Highly Significant ($P < 0.01$)

Fig. 2.4: Changes in Liver function tests of control and snake *Naja naja* venom treated albino rats.

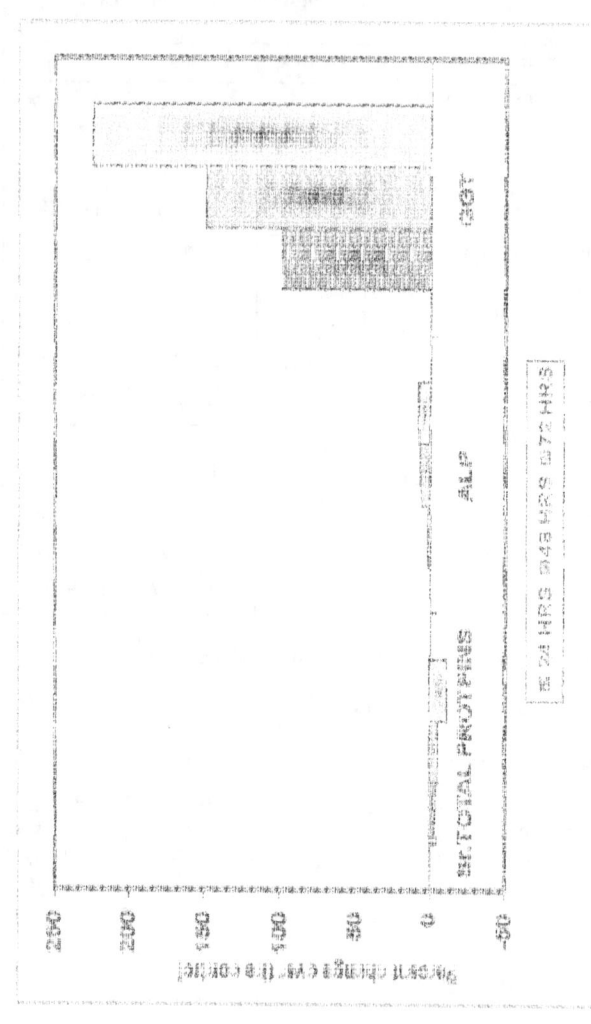

Table 2.5 : **Haemogram of Electrolytes of control and snake *Naja naja* venom treated albino rats.**

Parameters	Control	24 Hours	48 Hours	72 Hours
ELECTROLYTES **Sodium(mMol/L)** Mean SD PC	137.710 ±0.037	138.605 ±0.019 (0.650)	140.603 ±0.009 (2.100)	143.308 ±0.028 (4.065)
Potassium(mMol/L) Mean SD PC	5.505 ±0.017	5.936 ±0.013 (7.820)	6.678 ±0.028 (21.299)	7.745 ±0.019 (40.690)
Sr.Calcium(mMol/L) Mean SD PC	6.704 ±0.015	8.193 ±0.028 (22.203)	8.396 ±0.021 (25.236)	8.905 ±0.019 (32.831)
Sr.Phosphorous(mMol/L) Mean SD PC	21.998 ±0.009	21.996 ±0.015 (-0.007)	14.694 ±0.022 (-33.202)	10.696 ±0.017 (-51.379)
Sr.Amylase(IU/L) Mean SD PC	182.996 ±0.017	493.995 ±0.019 (169.949)	604.993 ±0.028 (230.605)	669.996 ±0.015 (266.127)
Lactate dehydrogenase(LDH) (IU/L) Mean SD PC	1850.008 ±0.028 (69.414	1670.996 ±0.015 (53.021)	1246.995 ±0.019 (382.59)	1092.006 ±0.022 (40.97)

All the values are mean ± SD of six individual observations.SD – Standard Deviation.
PC – Percent change over control.

ONE WAY ANOVA

Source of Variation	Df	Mean of Significance					
		Sodium	Potassium	Sr.Calcium	Sr.Phosphorous	Sr.Amylase	Lactate dehydrogenase
Between Groups	3	110.880**	17.313**	16.096**	567.127**	839240.844**	2263894.987**
Within Groups	20	0.013	0.008	0.009	0.005	0.008	0.009
Total	23						

NS : Not Significant, * Significant (P < 0.05),   ** Highly Significant (P < 0.01)

Fig.2.5 : Changes in Electrolytes of control and snake *Naja naja* venom treated albino rats.

Chapter-III

Protein metabolism

Chapter-III

A. Enzyme assay

Behavioral tolerance of individual animal to snake *Naja naja* venom influences several biochemical parameters, there by modifying the general metabolic state of the animal. Protein metabolism could be one of the major physiological events involved in the compensatory mechanism under stress condition.

Proteins

Proteins are the most abundant organic compounds and constitute a major part of the body dry weight (10-12 kg in adults). They perform a wide variety of static and dynamic functions. About half of the body protein is present in the supportive tissue while other half is intracellular. Proteins are large organic compounds made of amino acids arranged in a linear chain and joined together by peptide bonds between the carboxyl and amino groups of adjacent amino acid residues. Proteins are the ubiquitous macro molecules in the biological system and are derivatives of high molecular weight polypeptides (Murray *et al.*, 2007). They constitute about one fifth of an animal body on wet weight basis (Swaminathan, 1983). The concentration of proteins on tissue is a balance between the rate of their synthesis and degradation (Schmike, 1974). The overall protein turnover in animal is the dynamic equilibrium between synthesis and degradation rates (Tavill and Cooksly, 1983).

The literature pertaining to snake "*Naja naja* venom protein interaction" contributes immensely to the field of venom biochemistry. The venoms are found to alter the structural and soluble proteins by causing histopathological and biochemical changes in the cell (Shakoori *et al.*, 1976).

Proteins have important activities, including catalysis of metabolic reactions and transport of vitamins, minerals, oxygen and fuels. Some proteins make up the structure of tissues, while others function in nerve transmission, muscle contraction, cell motility and still others in blood clotting and immunologic defense, as hormones and regulatory molecules. Proteins are synthesized as a sequence of amino acids linked together in a linear polyamide (polypeptide) structure, but they assume complex three dimensional shapes in performing their function. There are approximately 300 amino acids present in various animal, plant and microbial systems, but only 20 amino acids are coded by DNA to appear in proteins. Many proteins also contain modified amino acids and accessory components, termed prosthetic groups. An important goal of molecular medicine is the identification of proteins whose presence, absence, or deficiency is associated with specific physiologic states or diseases (Murray *et al.*, 2007).

Proteins involve in cellular architecture, metabolic replications, enzyme mechanism. These are hydrolyzed to amino acids in the body which are further metabolized by incorporation into proteins or deamination or oxidation of amino acids (Murray *et al.*, 2007). Proteins play a dual role as a building material and as a source of energy for the organism (Babsky *et al.*, 1985). When there is no carbohydrate or fat availability, tissue proteins are the source of energy. In biological systems protein molecules may serve primary functions other than catalysis, the regulatory, signal transmission and recognition functions. Functionally proteins exhibit a great diversity and constitute heterogeneous group having diverse physiological functions as structural elements in contractile systems for nutrient storage, as vehicles of transport, as hormones, as catalysts, as toxins and as protective agents (Nelson and Cox, 2005).

The cellular proteins exist in soluble and insoluble form and the level of soluble protein is generally considered to be an index of aqueous state of cytosol which corresponds to the active metabolic state. Any alteration in protein metabolism is found to be the major physiological event in the compensatory mechanism in terms of homeostasis. The degradation of proteins was mainly brought about by protein hydrolyzing enzymes which cleave proteins into peptides and amino acids. A dynamic equilibrium exists between proteolysis and synthesis which is mainly responsible for protein turnover and homeostasis in any tissue (Grainde and Seglen, 1981).

The proportions of the different substances in venom and their specific characteristics vary among species. However, snakes with closer phylogenic relationship have more similarity in their venom properties and composition. The overall toxicity of the venom is due to enzymes as well as non enzymatic proteins (Tu, 1977). The complexity and composition of snake venoms and their great variability explain the huge diversity of their biological effects. The minor, non protein components of snake venom include inorganic and organic constituents. The organic constituents include free amino acids, peptides, nucleotides, carbohydrates (largely glycoproteins), lipids (phospholipids) and biogenic amines. The inorganic constituents include sodium, calcium, potassium, magnesium, zinc and small amounts of ferrous, cobalt and nickel.

Free Amino acids

Amino acids are called as building blocks; these are major components of proteins. Specificity of protein molecule is due to the number and sequence of amino acids. Most remarkable is that, cells can produce with strickingly different proteins and active by joining the same 20 amino acids in many different combinations and sequences (Nelson and Cox, 2005). In the protein molecule, the α–amino group is removed and the resulting skeleton is converted into major metabolic intermediate. Hydrolysis of dietary protein and endogenous protein result in the formation of amino acid pool in the body. The physiological state of the cell is depending upon its free amino acid pool (Vani, 1991). The protein constituents of plasma include a number of enzymes. Some of which are clinical diagnostic importance eg., amylase, lipase, phosphatases, amino transferases and glycolytic enzymes (Latner, 1975).

Amino acid metabolism is a complex system involving transamination, oxidation etc. Transamination is an important way of amino acid metabolism which is involved in transfer of amino group from one amino acid to another keto acid. Transaminases catalyze the transfer of α – amino acid to α- keto acid. These enzymes are called aminotransferases (Lehninger, 1995). The oxidation of amino acids is mediated by amino transferases. A total of 150 amino acid sequences of vitamin B6 dependent enzymes are known till date (Mehta *et al.,* 1993).

In addition to their roles as building blocks for peptides and proteins these are precursors of neurotransmitters and hormones. The carbon skeletons of some amino acids can be used to produce glucose through gluconeogenesis, there by providing a metabolic fuel for tissues that require or prefer glucose; such amino acids are designated as glucogenic-glycogenic amino acids. The carbon skeletons of some amino acids can also produce the equivalent of acetyl-CoA or acetoacetate and are termed ketogenic, indicating that they can be metabolized to give immediate precursors of lipids or ketone bodies. In an individual consuming adequate amount of protein, a significant quantity of amino acids may also be converted to carbohydrate (glycogen) or fat (triacylglycerols) for storage.

As the primary form in which the nitrogen is removed from amino acids is ammonia. When amino acids are metabolized, the resulting excess nitrogen must be excreted. Because free ammonia is quite toxic to humans and higher animals, rapidly convert the ammonia derived from amino acid catabolism to urea, which is less toxic, very soluble, and excreted in the urine. Thus the primary nitrogenous excretion product in humans is urea, produced by urea cycle in liver.

Proteases

Protease is an enzyme that conducts proteolysis, that begins protein catabolism by hydrolysis of the peptide bonds that link amino acids together in the polypeptide chain. Proteases are known to breakdown proteins to small peptides and ultimately to amino acids. They are present in almost all the tissues of mammals (Barret, 1977). Synthesis of macro molecules is a general phenomenon in all living organisms which involves a continuous turnover of cellular components.

The proteases with neutral pH is associated with peroxisomes and lysosomes referred as neutral proteases. Among the proteases, some are lysosomal in origin having acidic pH optimum, which are generally termed as cathepsins (Stagni and Debernard, 1968). Besides these two types, other type of protease with an alkaline pH optimum was detected in cytosolic fraction generally called as alkaline protease. Increase in acidic protease activity may be due to increase in number and size of lysosomes. Neutral proteases causing structural organization in different tissues and causes disassembly of intact myofibrils during metabolic turnover of myofibrillar proteins (Pellegrino and Franzini, 1963).

The changes in protease activities indicate the changes in energy cycle. All the proteins under normal conditions, irrespective of their location, are continuously degraded and replaced by new ones (Goldberg and Dice, 1974). Proteolytic activity is known to increase in various physiological and pathological conditions (Venkataswamy, 1991). The free amino acids and protease variations are, however species and tissue specific. Proteases are found to be altered under venom stress (Usha Rani, 2010).

Aspartate and Alanine aminotransferases (AST and ALAT)

The principal mechanism for removal of amino groups from the common amino acids is via transamination or transferring the amino group from amino acid to a suitable α-keto acid acceptor. Several enzymes, called aminotransferases (or transaminases) are capable of removing the amino group from most amino acids and producing the corresponding α-keto acid. Aminotransferase enzymes use pyridoxal phosphate, a cofactor derived from the vitamin B_6. In the catabolism of amino acids, aminotransferases play a dominant role. These are the key enzymes of nitrogen metabolism and in energy mobilization (Calabrese *et al.*, 1977). Aspartate and alanine aminotransferases are present both in mitochondria and cytosolic fractions of animal (Walton and Cowey, 1982).

Transaminases are important enzymes in animal metabolism which are intimately associated with amino acid synthesis and lysis. Among these, aspartate and alanine transaminases (AST and ALAT) are widely distributed in the cells of all animals. The AST catalyses the inter conversion of aspartic acid and α- ketoglutaric acid to oxalo acetic acid and glutamic acid. While ALAT catalyses the inter conversion of alanine and α-ketoglutaric acid to pyruvic acid and glutamic acid. The enzyme glutamate dehydrogenase plays a significant role in the catabolism of amino acids. It catalyses the reversible oxidative deamination of glutamate to α-ketoglutarate and ammonia with pyridine nucleotide (NAD or NADP) as coenzyme. All these enzymes functions as a link between protein and carbohydrate metabolisms and the net out come is incorporation of keto acids into the TCA cycle. There is much evidence for the alteration in the activities of these enzymes to a variety of environmental and physiological conditions (Knox and Greengard, 1965).

The activities of these aminotransferases were shown to be altered in tissues under several pathological conditions (Varshneya *et al.,* 1983; Paul *et al.,* 1984). Elevated AST and ALAT activities can be considered as an index of gluconeogenesis (Murray *et al.,* 2007; Usha Rani, 2010).

Glutamate dehydrogenase (GDH)

Glutamate dehydrogenase (GDH) is a regulatory enzyme known to check the deamination process to minimize the ammonia level and plays a significant role in the catabolism of amino acids. All these enzymes function as a link between protein and carbohydrate metabolisms and the net out come is the incorporation of keto acids into the TCA cycle. There is much evidence for the shifts in the activities of these enzymes to a variety of environmental and physiological conditions (Knox and Greengard, 1965).

Glutamate dehydrogenase enzyme is present in cytoplasm and mito-chondria. The cytoplasmic GDH recycle the cytoplasmic origin of ammonia and keeps up glutamate level for mitochondrial transport. Subsequently mitochondrial GDH supplies α-ketoglutarate to Kreb's cycle especially when the animal is in stress condition. GDH plays a crucial role in the nitrogen metabolism by functioning both in amino acid catabolism and their biosynthesis. GDH allows the incorporation of ammonia into α-ketoglutarate before being transferred by transamination to other α-keto acids (Murray *et al.,* 2007). Braunstein (1973) suggested that most of NH_3 required for urea synthesis from amino acid nitrogen was liberated via the GDH reaction.

Glutamate dehydrogenase (GDH) catalyzes the reversible oxidative deamination of glutamate to α-ketoglutarate and ammonia. This reaction is important in the linkage of nitrogen metabolism to carbohydrate metabolism via Kreb's cycle and it is the main pathway for the transformation of ammonia to α-amino group nitrogen (Sund *et al.*, 1977). The general pattern of GDH activity correlates inversely with the decreasing ammonia concentration (Oja *et al.*, 1966) suggesting the role of GDH in the production as well as ammonia detoxification. Thus, glutamine can serve as a buffer for ammonia utilization, as source ammonia, and as a carrier of amino groups. Because ammonia is quite toxic, a balance must be maintained between its production and utilization. Glutamate dehydrogenase reaction is reversible under physiological conditions if amino groups are required for amino acid and other biosynthetic processes.

GDH can be modified by the energy charge (ADP) and essential effectors in protein synthesis (Dieter *et al.*, 1981). This suggests in addition to ammonia, GDH may be involved in regulation of energy production and growth. In addition to the role of glutamate as a carrier of amino groups to GDH, glutamate serves as a precursor of glutamine, a process that consumes a molecule of ammonia. This is important because glutamine, along with alanine, is a key transporter of amino groups between various tissues and the liver, and is present in greater concentrations than most other amino acids in blood. The three forms of the same carbon skeleton, α-ketoglutarate, glutamate, and glutamine are interconverted via aminotransferases, glutamate dehydrogenase, glutamine synthetase and glutaminase.

Ammonia

Ammonia is constantly being liberated in the metabolism of amino acids and other nitrogenous compounds. Ammonia is important for the metabolic functions such as acid-base regulation and for the synthesis of purines, pyramidines and non-essential amino acids (Kvamme, 1983). Ammonia is toxic, particularly to the central nervous system (CNS). Catabolism of amino acids generates ammonia (NH_3) and ammonium ions (NH_4^+).

Most ammonia is detoxified at its site of formation, by amination of glutamate to glutamine, which is mainly derived from muscle and used as an energy source by enterocytes. The remaining nitrogen enters the portal vein either as ammonia or as alanine, both of which are used by the liver for the synthesis of urea. Although this is a normal detoxifying reaction in cells, when concentrations of ammonia are significantly increased, supplies of α-ketoglutarate in cells of the CNS may be depleted, resulting in inhibition of the TCA cycle and production of ATP. Removal of excess ammonia from the circulation is essential, since it is toxic to central nervous system and also interferes with peripheral nervous system (Banister *et al.*, 1988). Ammonia is produced endogenously in different tissues through deamination of amino acids via glutamate dehydrogenase and purine nucleotide cycle via AMP deaminase (Waarde and Kesbeke, 1982).

When ammonia concentration builds up in the blood and other biological fluids, ammonia diffuses into cells and across the blood/brain barrier. This increase in ammonia causes an increased synthesis of glutamate from α-ketoglutarate and increased synthesis of glutamine. There may be additional mechanisms accounting for the bizarre behavior observed in individuals with high blood concentrations of ammonia. In the nitrogen metabolism ammonia is a toxic end product to which the cells are sensitive and it is proportional to the integrity of work done (Bhargava, 1982; Babij *et al.*, 1983). Though a small fraction of body's amino acid, nitrogen is present as ammonia, its importance in clinical manifestation and metabolic regulation was recognized by several investigators. Because of its ready diffusibility, ammonia is found virtually in all tissues (Bessman, 1976).

In tissues and body fluids, ammonia concentrations are maintained over a narrow range. When this limit exceeds, it alarms the body with pathological symptoms (Sakaguchi *et al.*, 1981). High levels of ammonia not only cause injurious effect on tissues but also induce coma and death. Hence, an inherent homeostatic regulatory mechanism will come into operation. Tissues pick up this excess ammonia and convert it into less toxic substances. The pathophysiology includes the damage of mitochondria, structural breakdown of synaptic membrane, defective neuronal transport, mitochondrial swelling etc., (Gupta and Prabhakar Rao, 1980). Oxidative metabolism is impaired by increased lactate/ pyruvate ratio during hyper ammonia.

Urea

Catabolism of proteins usually results in the production of some of the unwanted nitrogenous end products like ammonia, urea, and uric acid. Ammonia toxicity posed a problem and a variety of adaptations have been observed among animal groups to dispose of the toxic ammonia (Murray *et al.,* 2007). The urea cycle also functions in removing excess bicarbonate, which are derived from oxidative metabolism and there by helping in regulating the acid-base balance. Urea production has an additional function in the maintenance of osmotic pressure and in the production of arginine and ornithine.

Nitrogen atoms are incorporated into urea exclusively from these two sources, which like amino acid catabolism to energy metabolism. Ammonia produced primarily from glutamate (via the glutamate dehydrogenase reaction) enters the urea cycle as carbamyl phosphate. The second nitrogen is contributed to urea by aspartic acid. Fumarate is formed in this process and may be recycled via the TCA cycle to oxaloacetate, which can accept another amino group to form aspartate or participate in either the TCA cycle or gluconeogenesis. Thus the funneling of amino groups from other amino acids into glutamate and aspartate provides the nitrogen for urea synthesis in a form appropriate for the urea cycle. The other pathways that lead to the release of amino groups through the action of amino acid oxidases make relatively minor contributions to the flow of amino groups from amino acid to urea.

The present investigation gives a brief understating on the effect of snake *Naja naja* venom on protein metabolism in albino rats. However, the reports in this account were made mostly by taking a few parameters of protein metabolism in some selected vital organs of albino rats exposed to snake *Naja naja* sub lethal doses. Further, the information of the above studies is unable to provide a clear concept on the effects of snake *Naja naja* venom on protein metabolism of albino rats. Hence, an attempt is made in the present study to document the effect of snake *Naja naja* venom on some aspects of protein metabolism in the selected vital organs of albino rat.

RESULTS

Total Proteins

The results of total protein content of the control and experimental albino rats under snake *Naja naja* venom administration are given in Table 3A.1 and Figure 3A.1. The experimental rats exposed to snake *Naja naja* venom showed statistically significant (P<0.01) decrease of total protein content in liver, kidney, brain, heart respectively. The decrease in total protein content was time dependent in snake *Naja naja* venom treated rats.

In experimental conditions, the tissues have shown decreased total protein content in liver (37.91%) followed by kidney (23.81%), brain (22.16%) and heart (20.80%) in 72 hrs of snake *Naja naja* venom administration. The maximum decrease was observed in 72 hrs of snake *Naja naja* venom administration followed by 48 hrs, and 24 hrs envenomated rats.

The lyotrophic series of total protein content decrement in 72 hrs of snake *Naja naja* venom treated rats is as follows:

Liver > Kidney >Brain > Heart

Free amino acids

The results of total free amino acid content of the control and experimental albino rats under venom intoxification are given in Table 3A.2 and Fig. 3A.2. The experimental rats exposed to snake *Naja naja* venom showed statistically significant (P<0.01) increase of free amino acid content in liver, kidney, brain and heart respectively. The increase in free amino acid content was time dependent in snake *Naja naja* venom treated rats.

In experimental conditions, the tissues have shown increased free amino acid content in kidney (22.04%), followed by brain (28.22%), heart (28.45%) and liver (34.87%) in 72 hrs of snake *Naja naja* venom administration. The time interval was maintained constant and gradual increase was observed in 24hrs, 48hrs and 72 hrs of snake *Naja naja* venom treated rats.

The lyotrophic series of total free amino acid content increment of snake *Naja naja* venom treated rats is as follows:

Kidney > Brain > Heart > Liver

Protease

The results of protease activity in the control and experimental albino rats under the study are given in Table 3A.3 and Fig.3A.3. The experimental rats exposed to snake *Naja naja* venom showed statistically significant (P<0.01) increase of protease activity in liver, kidney, brain and heart respectively. The increase in protease activity was time dependent in snake *Naja naja* venom treated rats.

In experimental conditions, the tissues have shown increased protease activity in kidney (17.479%) followed by brain (31.33%), heart (34.25%) and liver (35.43%) in 72 hrs. The gradual increase was observed in 24 hrs, 48 hrs and 72 hrs of snake *Naja naja* venom treated rats.

The lyotrophic series of protease activity increment in 72 hrs of snake *Naja naja* venom treated rats is as follows:

Kidney > Brain > Heart > Liver

Aspartate aminotransferase

The results of aspartate aminotransferase activity in the control and experimental albino rats under the study are given in Table 3A.4 and Fig. 3A.4. The experimental rats exposed to snake *Naja naja* venom showed statistically significant (P<0.01) increase of aspartate aminotransferase activity in liver, kidney, brain, heart respectively. The increase in aspartate aminotransferase activity was time dependent in snake *Naja naja* venom treated rats.

In experimental conditions the tissues have shown increased aspartate aminotransferase activity in kidney (21.40%) followed by liver (21.93%), brain (23.53%) and heart (29.98%) in 72 hrs. The gradual increase was observed in 24 hrs, 48 hrs and 72 hrs in snake *Naja naja* venom treated rats.

The lyotrophic series of aspartate aminotransferase activity increment in 72 hrs of snake *Naja naja* venom treated rats is as follows :

Kidney > Liver > Brain > Heart

Alanine aminotransferase

The results of alanine aminotransferase activity in the control and experimental albino rats under the study are given in Table 3A.5 and Figure 3A.5. The experimental rats exposed to snake *Naja naja* venom showed statistically significant (P<0.01) increase of alanine aminotransferase activity in liver, kidney, brain, heart. The increase in alanine aminotransferase activity was time dependent in snake *Naja naja* venom treated rats.

In experimental conditions, the tissues have shown increased alanine aminotransferase activity in kidney (17.80%), followed by brain (20.11%), liver (29.36 %) and heart (33.15%) in 72 hrs of snake *Naja naja* venom administration. The gradual increase was observed in 24 hrs, 48 hrs and 72 hrs of snake *Naja naja* venom treated rats.

The lyotrophic series of alanine aminotransferase activity increment in 72 hrs of snake *Naja naja* venom treated rats is as follows:

Kidney > Brain >Liver > Heart

Glutamate dehydrogenase

The results of glutamate dehydrogenase activity in the control and envenomated albino rats under the study are given in Table 3A.6 and Fig.3A.6.

The experimental rats exposed to snake *Naja naja* venom showed statistically significant (P<0.01) increase of glutamate dehydrogenase activity in liver, kidney, brain and heart respectively. The increase in glutamate dehydrogenase activity was time dependent in snake *Naja naja* venom treated rats.

In experimental conditions, the tissues have shown increased glutamate dehydrogenase activity in heart (16.45%) followed by kidney (38.93%), brain (39.67%) and liver (45.91%) s in 72 hrs. The gradual increase was observed in 24 hrs, 48 hrs and 72 hrs of snake *Naja naja* envenomated rats.

The lyotrophic series of glutamate dehydrogenase activity increment in 72 hrs of snake *Naja naja* venom treated rats is as follows.

Heart > Kidney > Brain > Liver

Ammonia

The results of ammonia content of the control and envenomated experimental albino rats under the study are given in Table 3A.7 and Fig. 3A.7. The rats exposed to snake *Naja naja* venom showed statistically significant (P<0.01) increase of ammonia content in liver, kidney, brain, heart respectively. The increase in ammonia content was time dependent in snake *Naja naja* venom envenomated rats.

In experimental conditions the tissues have shown increased ammonia content in liver (18.76%) followed by kidney (28.346%), heart (44.41%) and brain (51.68%) in 72 hrs of snake *Naja naja* venom administration.

The lyotrophic series of ammonia increment in 72 hrs of snake *Naja naja* venom treated rats is as follows.

Liver > Kidney > Heart > Brain

Urea

The results of urea content of the control and envenomated albino rats under the study are given in Table 3A.8 and Fig.3A.8. The experimental rats

86

exposed to snake *Naja naja* venom showed statistically significant (P<0.01) increase of urea content in liver and kidney, brain and heart respectively. The increase in urea content was time dependent in snake *Naja naja* envenomated rats.

In experimental conditions the tissues have shown increased urea content in liver (31.10%) followed by heart (44.41%), brain (51.689%) and kidney (114.58%) in 72 hrs of snake *Naja naja* venom administration. The gradual increase was observed in 24 hrs, 48 hrs and 72 hrs of snake *Naja naja* envenomated rats.

The lyotrophic series of urea content increment in 72 hrs of snake *Naja naja* venom treated rats is as follows.

Liver > Heart > Brain > Kidney

DISCUSSION

The results indicate changes in protein metabolism and associated enzyme systems after the administration of snake *Naja naja* venom in different tissues of albino rat. The physiological and biochemical activities in the albino rats were completely disturbed after the oral administration of snake *Naja naja* venom.

Rabie, *et al.*, (1972), reported changes in the enzymatic activities of mammalian tissues could be one of the mechanisms by which venomous snakes produce harm. The venom may either act by activating or inhibiting enzyme activities in the cell or destruction of the cell organelles with liberation of particular enzymes (Moustafa, *et al.*, 1974).

Total proteins

Catabolism of proteins and amino acids make a major contribution to the total energy production in albino rats. Proteins being involved in the architecture and physiology of the cell, they seem to occupy a key role in cell metabolism (Murray *et al.*, 2007). The depletion of total protein content observed in this investigation (Table 3A.1 and Figure 3A.1) can be correlated to this fact. Bradbury *et al.*, (1987) pointed out that the decreased protein content might also be attributed to the destruction or necrosis of cellular function and consequent impairment in protein synthetic machinery. Protein depletion in tissues may constitute a physiological mechanism and may play an important role of compensatory mechanism under snake *Naja naja* venom stress, to provide intermediates to the Kreb's cycle. It has also been reported that this trend of proteins was to enhance osmolarity to compensate osmoregulatory problems encountered due to the leakage of ions and other essential molecules during snake *Naja naja* venom toxicity (Rafat Yasmees, 1986).

Proteins being the most important organic constituents of organs, their role in the compensatory mechanisms of an animal can be accepted during stress conditions (Singaraju *et al.,* 1991). Shifts in proteins may ultimately lead to alterations in the entire protein metabolism of animals. In the present study, effects of snake *Naja naja* venom on the protein metabolism of the tissues of the rats exhibited time dependent changes in all the tissues exposed to the sub lethal doses of snake *Naja naja* venom indicating the breakdown of these proteins due to the venom toxic stress. Generally the breakdown of proteins dominates over synthesis under enhanced proteolytic activity (Murray *et al.,* 2007). It is evident in the present study that the breakdown of proteins is associated with the steep elevation in protease activity and free amino acid levels in the organs of the rats exposed to the sub lethal doses of snake *Naja naja* venom.

Mommsen and Walsh (1992) have estimated that oxidation of nitrogenous substances may account for 41% and 85% of total energy production from proteins and amino acid, respectively. Klassan (1991) reported that the depletion of protein suggests increased proteolysis and possible utilization of the products of their degradation for metabolic purposes. They may be fed into TCA cycle through aminotransferase system to cope up with excess demand of energy during the elimination of toxicants from the body. The depletion of protein level induces to diversification of energy to meet the impending energy demands during the toxic stress (Jagadeesan and Mathivanan, 1999).

The total protein content of the tissues such as liver, kidney, brain, heart generally decreased with envenomation. Such a decrease was significant in the kidney. The disturbance of renal function by the venom and the hemorrhage usually associated with snake bites are the acute factors for the observed hypoproteinemia. Acute renal failure has been reported with glomerular tubular intersticial and vascular lesions, and cortical necrosis following snake bites (Tilbury *et al.*, 1987). The increased vascular permeability due to the toxic action of the venom (Meier and Stocker 1991) contribute to the loss of protein in the tissues. Marsh *et al.*, (1997) suggested that viper venom might bring about a typical pattern of hemorrhage in tissues. Thus, it is likely that such hemorrhage might have contributed to the decrease in the tissue protein observed in the present work.

Abdel-Nabi *et al.*, (1997), Fahim *et al.*, (1998), Marsh *et al.*, (1997) reported that there is a reduction of total proteins in envenomated rats. The observed effects upon those parameters might suggest that the snake venom could have disturbed protein synthesis in hepatocytes due to cellular damage together with haemorrhages in liver, kidney, brain, heart leading to protein loss. Acute renal damage together with glomerular, tubular, interstitial and vascular lesions have been reported following snake bite (Tilbury *et al.*, 1987, Sant *et al.*, 1972, Chugh *et al.*, 1975, Sitprija *et al.*, 1977, Aung-Khin *et al.*, 1978, Sitprija *et al.*, 1982).

Meier *et al.*, (1991) observed that hemorrhages in vital organs together with increased vascular permeability were observed in the majority of viper and pit viper envenomation. Such increased vascular permeability, together with renal damage would further aggravate the accompanying hypo proteinemia and hypo albuminaemia.

Mohamed *et al.*, (1981) explained the measurements of tissue enzyme activities are important in assessing the state of the liver, kidney, brain, heart. Severe hepatocellular injuries, necrosis of hepatocytes and acute renal damage in kidney were observed in rats after *Echis carinatus* venom envenomation.

The data presented on total protein content indicates that a significant decrease occurs in total protein content in all tissues under snake *Naja naja* toxicity on time dependent manner. Decrease in total protein content suggests its metabolic utilization under snake *Naja naja* venom toxic stress.

It is clear from the foregoing account that overall decrease in total protein with abnormal rise in free amino acid pool might be explained in terms of accelerated proteolytic activity observed in different tissues during snake *Naja naja* venom toxicity.

Free amino acids (FAA)

The elevation in free amino acid content in the present investigation (Table 3A.2 and Fig.3A.2) is consistent with the decreased protein level, enhanced protease activity and transaminase activity during snake *Naja naja* venom exposure to 24 hrs, 48 hrs and 72 hrs. The alterations in free amino acids indicate, the condition of the tissue, and their increase or decrease might be considered as the operation of the stress phenomenon at the tissue level (Shakoori *et al.*, 1976). Free amino acids thus increased may be fed into the TCA cycle through aminotransferases, possibly to be utilized for energy production. Presumably the degradation of proteins has led to FAA accumulation. This higher level of FAA can also be attributed to the decreased utilization of amino acids and is also suggestive of its involvement in the maintenance of osmotic and acid base balance (Moorthy *et al.*, 1984).

Amino acids may not only act as precursors for the synthesis of essential proteins, but also contribute towards gluconeogenesis, glycogenesis and keto acid synthesis (Murray *et al.*, 2007). Free amino acids are known to act as osmotic and ionic agents (Jurss, 1980) in maintaining ionic equilibrium between external and internal media.

The increase in FAA content is a clear indication of step up proteolysis and fixation of ammonia into keto acids resulting in amino acid synthesis. These two processes in general contribute to the amino acid pool. Thus, these amino acids may be utilized for the synthesis of new types of essential proteins and enzymes to cope with the toxic stress conditions to which the animal was exposed (James *et al.,* 1982). Elevated levels of free amino acids were observed in different tissues of albino rats exposed to certain pesticides (Usha Rani, 2010).

Free amino acids were elevated in all the tissues exposed to snake *Naja naja* venom. The elevation in free amino acids was in consonance with the increased proteolytic activity. The elevated free amino acid levels indicate altered protein homeostasis and nitrogen inbalance due to snake *Naja naja* venom toxicity.

Protease

Under proteolysis, enhanced breakdown dominates over synthesis. While in the case of anabolic process, increased synthesis dominates the protein breakdown (Murray *et al.,* 2007). Moreover, histopathological damage and hydro mineral imbalance during pesticide stress has been reported to account for the elevated protease activity (Moorthy *et al.,* 1984). Increase in protease activity observed at 24 hrs, 48 hrs and 72 hrs snake *Naja naja* envenomation in different tissues of albino rats were clearly reflected in breakdown of proteins (Table 3A.3 and Fig. 3A.3).

Proteases were found to be activated during stress condition indicating a possible relation between inactivation of oxidative enzymes, reduction in energy production and acceleration of proteolysis. Increase in protease activity in different tissues in the present study clearly reflected in the decrease of total protein levels in the tissues.

The elevated protease activity, in general, indicates profound loss of proteins causing structural disorganization and disassembly of structural proteins in different tissues during snake *Naja naja* envenomation.

Aspartate and Alanine aminotransferases (AST and ALAT)

Increased levels of AST and ALAT activities were observed (Tables 3A.4 and 3A.5 Fig. 3A.4 and 3A.5). This will give a clear indication of shunting of amino acids into TCA cycle through oxidative deamination and active transamination. Such a phenomena was necessary to cope up with the energy crisis during snake *Naja naja* venom stress. It has been suggested that stress conditions in general induce elevation in the transamination pathway (Awasthi *et al.*, 1984).

Increased AST and ALAT activities may be due to disruption of mitochondrial integrity or increased synthesis of enzymes. Increased ammonia content was also shown to be responsible for the increased transaminases activity. The aminotransferases play a very important role in the animal metabolism in the sense that they are intimately associated with amino acid synthesis and lysis. Aminotransferases play an important role in the utilization of amino acids for the oxidation of gluconeogenesis (Murray *et al.*, 2007). The depletion in energy reserve necessitates enhancement of other alternative mechanisms like aminotransferase reaction to feed the keto acids into the TCA cycle (Kabeer *et al.*, 1978).

The aspartate and alanine aminotransferases which function as a strategic link between carbohydrate and protein metabolisms are known to alter under severe pathological conditions, hence they are considered as sensitive indicators of stress. The aspartate and alanine aminotransferases are referred as "markers of cell injury" (Loeb, 1982) and are excellent indicators of early hepatic lesions, since they are first to leak out from the cell in the case of injury.

Increase in AST and ALAT in different tissues of the rats at different doses of snake *Naja naja* venom indicate the initial effort taken by the animal for raising its energy resources through active transamination and for the synthesis of new proteins required for detoxification of the toxicant and its disposal. To have an insight into the role of these enzymes in the altered metabolism of pesticide intoxificated rats the activities of both AST and ALAT were investigated in the

92

present experiment. Elevated levels of AST and ALAT indicate the enhanced transamination of amino acids, which may provide keto acids to serve as precursors in the synthesis of essential organic elements.

Activities of superoxide dismutase, catalase, glutathione and malonyl dialdehyde levels in the liver reflect the oxidative status and the serum enzymes like AST, ALAT and ALP represent the functional status of the liver (Manna *et al.,* 2005). Mukhopadhyay *et al.,* (1982) reported increased GOT and GPT activities in liver and serum of the carbofuran treated *Clarias batrachus* are compatible with liver damage and growth rate reduction.

Mohammed *et al.,* (1981) reported the same increase in serum AST, ALAT after using the venom of *Bitis arientans* and *Bitis gabonica.* Same sort of the results were observed in viper snakes *Echis carinatus* and *Cerastes cerastes.* In fact the total protein content of the hepatic and extra hepatic tissues were observed to decline severely after envenomation in our experiment, a protein could be the source of the gluconeogenic mechanism.

Rahmy, *et al.,* (1995) reported increase in liver, kidney, brain, heart have significant amount of ALAT, because it is more specific to liver cells. Moreover, increased AST levels as result of Elapidae neurotoxins and cardiotoxins effects were found to induce severe myonecrosis and fatal myocardial injury. The increase in ALP activity in snake envenomated rats might be attributed to the destruction of liver cells (Abdel-Nabi *et al.,* 1993).

The activity levels of aminotransferases (AST and ALAT) were elevated in all the tissues studied. The elevated activity indicate the mobilization of free amino acids towards gluconeogenesis to meet the energy demands during snake *Naja naja* venom toxicity. Elevated levels of AST and ALAT indicate the enhanced transamination of amino acids which may provide keto acids to serve as precursors in the synthesis of essential organic constituents.

Glutamate dehydrogenase (GDH)

GDH catalyzes the reversible reaction of oxidative deamination of glutamate to α-ketoglutarate and ammonia and plays an important role in the catabolism and biosynthesis of amino acid (Murray *et al.,* 2007). GDH activity levels were increased in the tissues of rats exposed to sub lethal dose of snake *Naja naja* venom (Table 3A.6 and Fig. 3A.6).

The elevation observed in the GDH activity indicates its contribution to enhanced ammonia levels and glutamate oxidation during snake *Naja naja* venom toxicity. Increased free amino acid levels and their subsequent transamination results in greater production of glutamate, thus increasing the intracellular availability of substrate, glutamate for consequent oxidative deamination reaction through GDH. Besides the elevation of transaminases and GDH helps in supplying keto acids to the TCA cycle in order to compensate the energy crisis in different tissues during snake *Naja naja* envenomation.

Glutamate dehydrogenase occurs with high activity in the mitochondrial matrix and is commonly used as a marker for matrix space (Kovacevic and Mc Givan, 1983). It has a great importance in neuro transmitter balance in brain tissue and maintenance of nitrogen in liver tissue. As GDH plays an important role in detoxification of ammonia (Campbell, 1973), increased glutamate dehydrogenase activity was observed in the tissues of albino rat exposed to snake *Naja naja* venom.

Glutamate dehydrogenase (GDH) play a crucial role in the cells affected by a variety of effectors of protein metabolism in the cells (Ramanadikshithulu *et al.,* 1976). This enzyme has several metabolic functions with great physiological significance and closely associated with the detoxification mechanisms of tissues. GDH in extra-hepatic tissues could be utilized for channeling of ammonia released during proteolysis for its detoxification into urea in the liver. Hence, the activities of AST, ALAT and GDH are considered as sensitive indicators of stress (Gould *et al.,* 1976).

In the present study, increase in GDH activity, AST and ALAT favors trans-deamination of amino acids to incorporate them into TCA cycle for energy releasing purposes to meet the imposed toxic stress as keto acids (Ramana-dikshithulu *et al.*, 1976; Sreedevi *et al.*, 1992). A progressive increase in the levels of ammonia in major vital organs of the animal exposed to pesticide at different doses. The elevation in GDH activity under toxic stress was also reported (Radha krishnaiah *et al.*, 1991; Sreedevi *et al.*, 1992).

The GDH activity was found to be elevated in all the tissues and the elevated GDH activity levels indicate its contribution to ammonia production and glutamate oxidation during snake *Naja naja* envenomation. The elevated free amino acid levels and their subsequent transamination towards the formation of glutamate leads to the consequent oxidative deamination reaction through GDH and also helps in supplying keto acids to TCA cycle for energy production.

Ammonia

In the present study, ammonia content is increased in the tissues of rats exposed to $1/50^{th}$ sub lethal dose of snake *Naja naja* envenomation (Table 3A.7 and Fig.3A.7).

Elevated activities of proteases, transaminase reactions, deamination reaction (GDH) support the ammonia levels in different tissues of snake *Naja naja* venom treated rats. The enhanced ammonia levels in different tissues of snake *Naja naja* venom treated rats may lead to aminotoxaemia, and show deleterious effects on the animal metabolism. Though ammonia is essential for the synthesis of important compounds such as purines, pyramidines and non-essential amino acids, it also play a key factor in acid-base regulation and is toxic in non-physiological concentrations and excess ammonia therefore has to be disposed off (Murray *et al.*, 2007).

In the present study elevated activities of protease, transaminase reactions and increased deamination reactions (GDH) support the increase in ammonia levels during snake *Naja naja* envenomation.

Urea

The elevation in urea levels was in consonance with increased proteolytic activity, enhanced transamination and elevated ammonia levels during snake *Naja naja* venom toxicosis. Increased levels of urea under snake *Naja naja* venom stress reveal that the rats might have adapted to the biosynthesis of urea as a major pathway of detoxification of ammonia. Probably this pathway may be beneficial to animals in detoxification and physiological compensation or adjustment to various exogenous and endogenous toxicants.

The blood urea level in viper bite cases increased significantly after the sixth hour. Since anti-venom does not decrease the blood urea to normal, dialysis is required for normalization of urea level (Pradeep kumar and Basheer, 2011).

Al Jammaz (1994) studied the effect of *Walterinnesia aegyptia* and *Echis coloratus* venom on solute levels in the plasma of albino rats and observed a rise in plasma and urea level. This has been observed in rabbits injected with scorpion venoms, (Ismail, 1978) and could probably be due to the impairment of renal function, (Gitter, 1962). Such rise might explain the increased plasma osmolarity. However plasma chloride level did not vary significantly with venom of both snakes used. This might reflect the stability of chloride reabsorption by the kidneys of rat, since sodium and chloride ions do not always move together in the nephron.

In the present study, the elevation in urea levels was in consonance with increased proteolytic activity, enhanced transamination and elevated levels of ammonia during snake *Naja naja* venom intoxification. Number of reports are available on the impact of different groups of pesticides on protein metabolism of non-target organisms (Oruc and Uner, 1999).

Faster accumulation of pesticide in the active tissues of the animal and progressive decrease of oxidative metabolism might have not aided them to overcome the imposed toxic stress. Degradation of proteins results in increased amino acid pool. Prevalence of pathological conditions in an animal also decreases protein synthetic potentials which in turn increase the amino acid pool. There are reports indicating that increased amino acids may be used either for the synthesis of required proteins (James *et al.,* 1979) or for the production of energy (Lowenstein, 1972) to cope up with the toxic stress. Increase in free amino acids could be partly due to increased proteolysis and partly be due to stepping of their synthesis.

Despite the toxic effects of snake *Naja naja* venom exposure to sub lethal doses (24 hrs, 48hrs and 72hrs) the rats tried to withstand the toxic effects imposed by the venom in modulating their physiological and metabolic response towards proper utilization of proteins, free amino acids for synthetic processes.

The observations made in the rats under snake *Naja naja* venom intoxification, it is concluded that the changes are dependent on the time interval. Highest time interval (i.e., 72 hrs) caused more damage to the physiological, biochemical activities in the rats. Hence, there appeared an irrecoverable loss to the biochemical integrity of the cells due to snake *Naja naja* venom. The alterations in protein metabolism were also observed at lowest time interval (i.e., 24hrs) as the stress prevailed in lowest time.

In the present investigation snake *Naja naja* venom has altered some enzymes of protein metabolism in experimental rats. However, tissues have shown highly significant changes in all the parameters investigated.

Table 3A.1: Changes in total protein content (mg/gm wet wt of tissue) in different tissues of control and snake *Naja naja* treated albino rats. Values in parentheses indicate percent change over control

Name of the tissue	Control	24 Hours	48 Hours	72 Hours
Liver				
Mean	160.964	140.432	117.876	99.942
SD	±1.878	±0.776	±0.664	±0.218
PC		(-12.756)	(-26.769)	(-37.911)
Kidney				
Mean	118.367	109.759	99.548	90.181
SD	±0.737	±0.896	±0.609	±0.494
PC		(-7.272)	(-15.899)	(-23.812)
Brain				
Mean	115.189	100.432	92.663	89.663
SD	±0.726	±0.695	±0.466	±0.833
PC		(-12.811)	(-19.556)	(-22.160)
Heart				
Mean	97.883	88.533	80.505	77.515
SD	±0.485	±0.296	±0.642	±0.334
PC		(-9.552)	(-17.754)	(-20.809)

All the values are mean ± SD of six individual observations.
SD – Standard Deviation.
PC – Percent change over control.

ONE WAY ANOVA

Source of Variation	DF	Liver	Kidney	Brain	Heart
		MS	MS	MS	MS
Between Groups	3	12707.586**	2696.992**	2343.131**	1498.582**
Within Groups	20	23.090	9.813	9.601	4.231
Total	23				

NS: Not Significant, *-Significant (P<0.05), **- Highly Significant (P<0.01)

Fig. 3A.1 : Total protein content in different tissues of albino rats exposed to snake *Naja naja* venom.

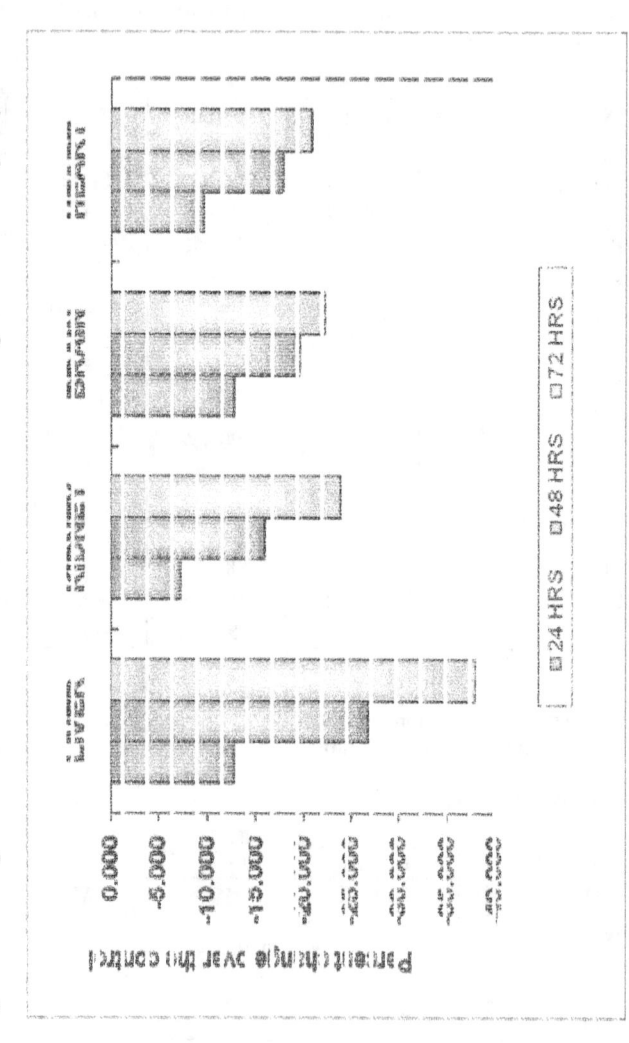

Table 3A.2: **Changes in free amino acid content (μ moles of tyrosine/gm wet wt of tissue) in different tissues of control and snake *Naja naja* venom treated albino rats. Values in parentheses indicate percent change over control.**

Name of the tissue	Control	24 Hours	48 Hours	72 Hours
Liver				
Mean	11.133	12.751	13.122	15.016
SD	±0.011	± 0.012	±0.013	±0.014
PC		(14.533)	(17.863)	(34.875)
Kidney				
Mean	7.748	8.349	8.967	9.456
SD	± 0.010	±0.011	±0.014	±0.021
PC		(7.752)	(15.722)	(22.043)
Brain				
Mean	5.448	5.845	6.745	6.985
SD	± 0.013	±0.014	± 0.013	±0.012
PC		(7.303)	(23.821)	(28.227)
Heart				
Mean	3.987	4.121	4.746	5.122
SD	± 0.016	± 0.024	± 0.013	±0.016
PC		(3.365)	(19.024)	(28.458)

All the values are mean ± SD of six individual observations.
SD – Standard Deviation.
PC – Percent change over control.

ONE WAY ANOVA

Source of Variation	DF	Liver	Kidney	Brain	Heart
		MS	MS	MS	MS
Between Groups	3	45.752	9.914**	9.560**	5.120**
Within Groups	20	0.003	0.004	0.003	0.006
Total	23				

NS: Not Significant, *-Significant ($P<0.05$), **- Highly Significant ($P<0.01$)

Fig. 3A.2 : Free amino acid content in different tissues of albino rats exposed to snake *Naja naja* venom.

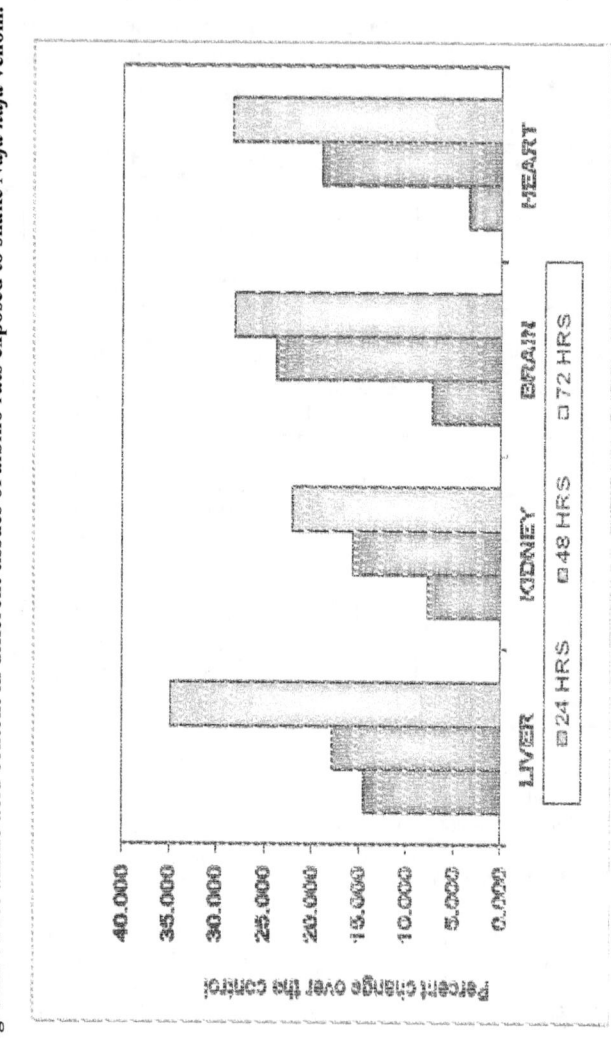

Table 3A.3: Changes in protease activity (μ moles of tyrosine/mg protein/hr) levels in different tissues of control and snake *Naja naja* venom treated albino rats. Values in parentheses indicate percent change over control.

Name of the tissue	Control	24 Hours	48 Hours	72 Hours
Liver				
Mean	1.237	1.275	1.473	1.676
SD	±0.012	±0.014	±0.015	±0.027
PC		(3.031)	(19.060)	(35.439)
Kidney				
Mean	0.528	0.552	0.603	0.621
SD	±0.014	±0.017	±0.021	±0.023
PC		(4.385)	(14.164)	(17.476)
Brain				
Mean	0.385	0.423	0.452	0.505
SD	±0.020	±0.027	±0.029	±0.073
PC		(10.100)	(17.555)	(31.339)
Heart				
Mean	0.312	0.334	0.372	0.419
SD	±0.031	±0.035	±0.040	±0.042
PC		(6.990)	(19.157)	(34.258)

All the values are mean ± SD of six individual observations.
SD – Standard Deviation.
PC – Percent change over control.

ONE WAY ANOVA

Source of Variation	DF	Heart	Liver	Kidney	Muscle
		MS	MS	MS	MS
Between Groups	3	0.736**	0.034**	0.046**	0.040**
Within Groups	20	0.007	0.007	0.037	0.028
Total	23				

NS: Not Significant, *-Significant ($P<0.05$), **- Highly Significant ($P<0.01$)

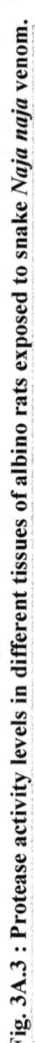

Fig. 3A.3 : Protease activity levels in different tissues of albino rats exposed to snake *Naja naja* venom.

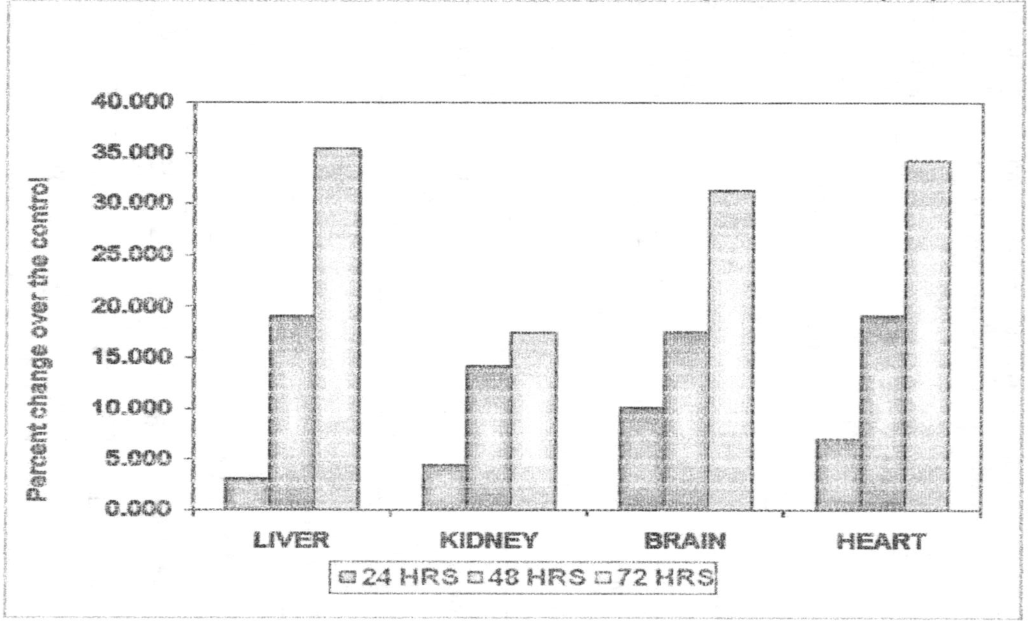

Table 3A.4: Changes in aspartate aminotransferase (μ moles of pyruvate formed/mg protein/hr) levels in different tissues of control and snake *Naja naja venom* treated albino rats. Values in parentheses indicate percent change over control.

Name of the tissue	Control	24 Hours	48 Hours	72 Hours
Liver				
Mean	1.547	1.641	1.733	1.886
SD	±0.012	±0.011	±0.014	±0.013
PC		(6.055)	(12.034)	(21.935)
Kidney				
Mean	0.228	0.247	0.264	0.277
SD	±0.014	±0.013	±0.011	±0.011
PC		(8.108)	(15.632)	(21.402)
Brain				
Mean	0.501	0.538	0.571	0.619
SD	±0.010	±0.014	±0.009	±0.013
PC		(7.490)	(14.048)	(23.535)
Heart				
Mean	0.575	0.623	0.701	0.747
SD	±0.008	±0.009	±0.011	±0.010
PC		(8.350)	(22.006)	(29.980)

All the values are mean ± SD of six individual observations.
SD – Standard Deviation.
PC – Percent change over control.

ONE WAY ANOVA

Source of Variation	DF	Liver MS	Kidney MS	Brain MS	Heart MS
Between Groups	3	0.376**	0.008**	0.045**	0.108**
Within Groups	20	0.003	0.003	0.003	0.002
Total	23				

NS: Not Significant, *-Significant ($P<0.05$), **- Highly Significant ($P<0.01$)

Fig. 3A.4 : Aspartate aminotransferase activity levels in different tissues of albino rats exposed to snake *Naja naja* venom.

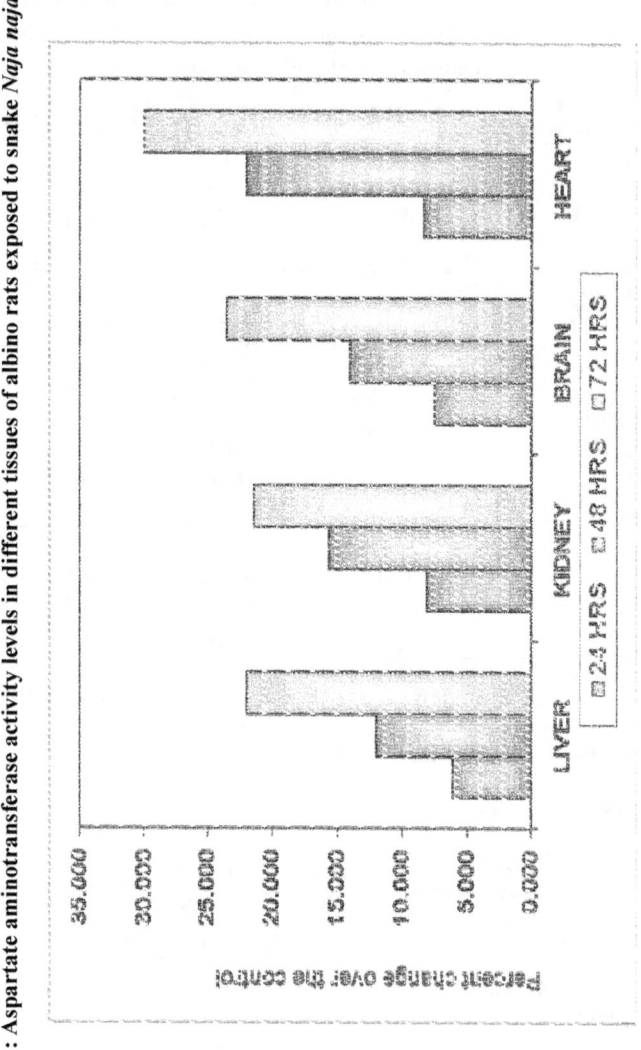

Table 3A.5: Changes in alanine aminotransferase (μ moles of pyruvate formed/mg protein/hr) levels in different tissues of control and snake *Naja naja* venom treated albino rats. Values in parentheses indicate percent change over control.

Name of the tissue	Control	24 Hours	48 Hours	72 Hours
Liver				
Mean	1.811	1.847	1.948	2.343
SD	±0.014	±0.013	±0.012	±0.010
PC		(1.960)	(7.537)	(29.364)
Kidney				
Mean	0.846	0.873	0.922	0.997
SD	±0.013	±0.012	±0.014	±0.012
PC		(3.191)	(8.983)	(17.809)
Brain				
Mean	0.746	0.775	0.806	0.896
SD	±0.014	±0.013	±0.014	±0.013
PC		(3.979)	(8.114)	(20.116)
Heart				
Mean	0.487	0.527	0.568	0.649
SD	±0.009	±0.010	±0.008	±0.008
PC		(8.108)	(16.593)	(33.151)

All the values are mean ± SD of six individual observations.
SD – Standard Deviation.
PC – Percent change over control.

ONE WAY ANOVA

Source of Variation	DF	Liver MS	Kidney MS	Brain MS	Heart MS
Between Groups	3	1.073**	0.079**	0.076**	0.086**A
Within Groups	20	0.003	0.003	0.004	0.002
Total	23				

NS: Not Significant, *-Significant (P<0.05), **- Highly Significant (P<0.01)

Fig. 3A.5 : Alanine aminotransferase activity levels in different tissues of albino rats exposed to snake *Naja naja* venom.

Table 3A.6: Changes in glutamate dehydrogenase (GDH) (μ moles of formazon formed/mg protein/hr) levels in different tissues of control and snake *Naja naja* venom treated albino rats. Values in parentheses indicate percent change over control.

Name of the tissue	Control	24 Hours	48 Hours	72 Hours
Liver				
Mean	0.249	0.274	0.297	0.363
SD	±0.012	±0.016	±0.017	±0.106
PC		(10.174)	(19.076)	(45.917)
Kidney				
Mean	0.219	0.235	0.283	0.305
SD	±0.015	±0.017	±0.007	±0.016
PC		(7.148)	(29.049)	(38.935)
Brain				
Mean	0.210	0.230	0.283	0.293
SD	±0.013	±0.011	±0.026	±0.024
PC		(9.548)	(34.987)	(39.677)
Heart				
Mean	0.116	0.117	0.123	0.135
SD	±0.011	±0.013	±0.013	±0.012
PC		(3.100)	(6.349)	(16.450)

All the values are mean ± SD of six individual observations.
SD – Standard Deviation.
PC – Percent change over control.

ONE WAY ANOVA

Source of Variation	DF	Heart	Liver	Kidney	Muscle
		MS	MS	MS	MS
Between Groups	3	0.043**	0.029**	0.029**	0.001*
Within Groups	20	0.060	0.004	0.008	0.003
Total	23				

NS: Not Significant, *-Significant ($P<0.05$), **- Highly Significant ($P<0.01$)

Fig. 3A.6 : Glutamate dehydrogenase activity levels in different tissues of albino rats exposed to snake *Naja naja* venom.

Table 3A.7: Changes in ammonia levels (μ moles of ammonia/gm wet wt of tissue) in different tissues of control and snake *Naja naja* venom treated albino rats. Values in parentheses indicate percent change over control.

Name of the tissue	Control	24 Hours	48 Hours	72 Hours
Liver				
Mean	11.143	11.752	12.129	12.988
SD	±1.428	±0.009	±0.019	±0.007
PC		(7.471)	(10.917)	(18.769)
Kidney				
Mean	3.987	4.117	4.124	5.117
SD	±0.009	±0.008	± 0.007	± 0.006
PC		(3.269)	(69.819)	(28.346)
Brain				
Mean	0.163	0.167	0.197	0.247
SD	±0.018	±0.017	±0.034	±0.017
PC		(2.559)	(20.676)	(51.689)
Heart				
Mean	0.116	0.128	0.162	0.168
SD	±0.007	±0.007	±0.007	±0.007
PC		(10.029)	(39.255)	(44.413)

All the values are mean ± SD of six individual observations.
SD – Standard Deviation.
PC – Percent change over control.

ONE WAY ANOVA

Source of Variation	DF	Liver	Kidney	Brain	Heart
		MS	MS	MS	MS
Between Groups	3	10.731**	4.949**	0.027**	0.012**
Within Groups	20	10.203	0.001	0.010	0.001
Total	23				

NS: Not Significant, *-Significant (P<0.05), **- Highly Significant (P<0.01)

Fig. 3A.7 : Ammonia levels in different tissues of albino rats exposed to snake *Naja naja* venom.

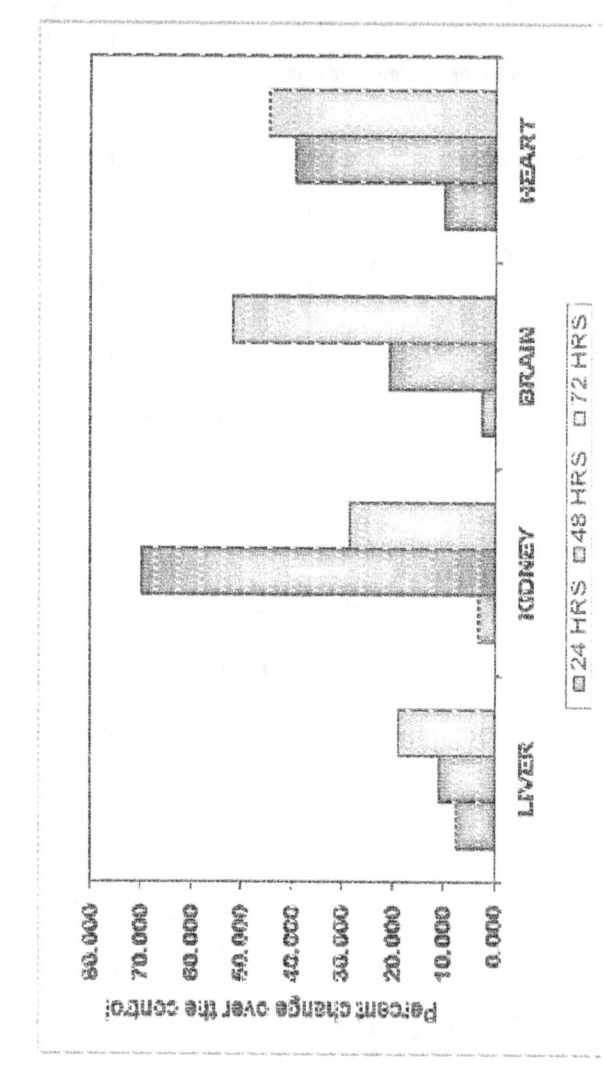

Table 3A.8: Changes in urea levels (μ moles of urea /gm wet wt of tissue) in different tissues of control and snake *Naja naja* venom treated albino rats. Values in parentheses indicate percent change over control.

Name of the tissue	Control	24 Hours	48 Hours	72 Hours
Liver				
Mean	3.144	3.648	3.848	4.122
SD	± 0.014	± 0.019	± 0.015	± 0.015
PC		(16.037)	(22.404)	(31.109)
Kidney				
Mean	0.977	1.250	1.839	2.096
SD	± 0.016	±0.018	±0.025	±0.018
PC		(27.982)	(88.296)	(114.588)
Brain				
Mean	0.163	0.165	0.197	0.247
SD	±0.018	±0.015	±0.034	±0.017
PC		(4.433)	(20.676)	(51.689)
Heart				
Mean	0.116	0.128	0.162	0.168
SD	±0.007	±0.007	±0.007	±0.007
PC		(10.029)	(39.255)	(44.413)

All the values are mean ± SD of six individual observations.
SD – Standard Deviation.
PC – Percent change over control.

ONE WAY ANOVA

Source of Variation	DF	Liver	Kidney	Brain	Heart
		MS	MS	MS	MS
Between Groups	3	3.069**	4.800**	0.028**	0.012**
Within Groups	20	0.005	0.008	0.010	0.001
Total	23				

NS: Not Significant, *-Significant (P<0.05), **- Highly Significant (P<0.01)

Fig. 3A.8 : Urea levels in different tissues of albino rats exposed to snake *Naja naja* venom.

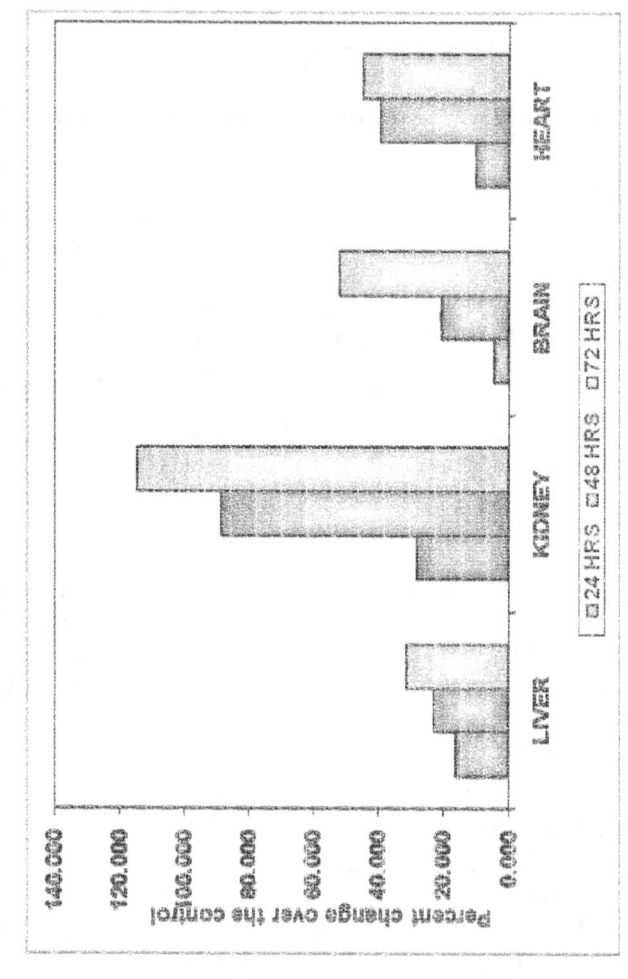

Chapter-III

B. Identification of DNA Damage

3B. Agarose Gel Electrophoresis

Electrophoresis is a technique in which the migration of charged particles moves under the influence of electric field. Many biological molecules, such as amino acids, proteins, peptides, nucleotides and nucleic acids possess ionisable groups and therefore at any given pH solution as electrically charged species either cations or anions are under the influence of electric field.

The gel is made from agarose, a highly purified form of the polysaccharide that is used to make agar plates. The gel is immersed in buffer and the DNA fragments are loaded onto a well at one end of the gel and made to move through the gel by the application of electric current. DNA is negatively charged and so will move through faster and these fragments separate according to size.

The DNA bands are visualized by adding ethidium bromide (EtBr), a fluorescent molecule which intercalate with the DNA bases, extending the length of linear and nicked circular DNA molecules and making them more rigid. When EtBr is added, UV radiations at 254 nm is absorbed by the DNA and transmitted to the bound eye. EtBr is a powerful mutagen and hence the gel should be handled carefully with gloves. The DNA bands can be visualized under UV and the data can be recorded by gel documentation appliances.

Characteristic features of gel electrophoresis are:

1. (i)   **The molecular weight of the DNA**: The migration rate is inversely proportional to the molecular weight.

     (ii)   **Agarose concentration**: The migration rate is inversely propor-tional to the agarose concentration.

     (iii)   **Conformation of the DNA:** Linear form travels slowest and the super coiled form travels fastest.

     (iv)   **Applied voltage**: Typical value 5 volts per cm. The heat generated during electrophoresis is dissipated by the buffer.

115

2.   DNA being a polyanionic at neutral pH, it migrates towards the anode.

3.   The loading dye for DNA contains glycerol, which gives density to help the sample sink to the bottom of the well and marker dyes Xylenecyanol and Bromophenol blue. Bromophenol blue on par with 300-400 bp DNA and xylenecyanol with 2-3 bp DNA.

4.   The DNA is visualized by adding EtBr a fluorescent molecule which intercales with DNA bases. To 0.8% agarose gel add EtBr to give 0.5 µg/ml concentration. UV radiation at 254 nm is absorbed by the DNA and transmitted to the bound eye. The energy is re-emitted at 590 nm in the red orange region of the spectrum.

5.   EtBr is a powerful mutagen. The dye is usually incorporated into the gel or conversely the gel is stained after running by soaking in a solution of EtBr.

6.   The usual sensitivity of detection is 0.1 µg of DNA.

7.   The gel will be run along with the molecular weight marker, a wide range of which is commercially available.

Results

In the present investigation the DNA samples of control and   snake *Naja naja* venom treated albino rat liver, kidney, brain, heart were used to run the Agarose gel electrophoresis to know the DNA damage pattern along with molecular marker (Plate 3B; Fig.3B; Table 3B).

Snake *Naja naja* venom affect the major vital organs like liver, kidney, brain, heart and the DNA banding pattern was found to be altered in envenomated rats. The DNA banding pattern revealed that the snake *Naja naja* venom exposed liver, kidney, brain, heart interact with the DNA intact bands, and there by DNA lysis takes place causing DNA damage indicating that there is a significant DNA fragmentation on time dependent manner i.e., 24 hrs, 48 hrs, and 72 hrs seen during the experimental condition.

In liver, kidney, brain, heart tissues of snake venom envenomated rats, DNA fragmentation was observed. The DNA fragmentation gradually increased from 24 hrs, 48 hrs, 72 hrs (Plate 3B; Fig. 3B; Table 3B).

Discussion

The toxicity of a compound depends upon the degree of severity of uptake, distribution and metabolism, as well as its molecular interaction at the site of action and mode of administration. Broad *et al.*, (1979) demonstrated by two-directional polyacrylamide gel electrophoresis that venom from the two snakes had characteristically distinct electrophoretograms and this has been used a tool to distinguish the species of different snakes in taxonomy.

The electrophoretic DNA patterns are consistent in 24 hrs, 48 hrs, 72 hrs of envenomated rats. The present study may be for the first time concentrated on DNA banding pattern in envenomated rats and this banding pattern will have the significance in building up the different snake species in getting together under one umbrella of taxonomy.

Table 3B: **Changes in DNA fragmentation analysis of Liver, Kidney, Brain, Heart of albino rats in 24 hrs, 48hrs, 72 hrs exposed to snake** *Naja naja* **Venom. Values in parentheses indicate percent change over control**

Name of the tissue	Control	24 Hours	48 Hours	72 Hours
Liver				
Mean	8.917	14.833	17.833	24.833
SD	±2.010	±2.090	±2.041	±1.992
PC		(66.355)	(100.00)	(178.505)
Kidney				
Mean	7.833	13.833	16.583	20.750
SD	± 2.090	±2.090	±2.333	±2.115
PC		(76.596)	(111.702)	(164.894)
Brain				
Mean	5.750	9.833	13.500	18.750
SD	±1.969	±2.090	±2.470	±2.115
PC		(71.014)	(134.783)	(226.087)
Heart				
Mean	4.750	7.833	11.500	16.750
SD	±2.043	±2.090	±2.470	±2.115
PC		(64.912)	(142.105)	(252.632)

All the values are mean ± SD of six individual observations.
SD – Standard Deviation.
PC – Percent change over control.

ONE WAY ANOVA

Source of Variation	DF	Liver	Kidney	Brain	Heart
		MS	MS	MS	MS
Between Groups	3	262.927**	176.083**	183.125**	159.792**
Within Groups	20	4.135	4.663	4.704	4.779
Total	23				

NS: Not Significant, *-Significant (P<0.05), **- Highly Significant (P<0.01).

Fig.3B : DNA Fragmentation analysis of Liver, Kidney, Brain and Heart of albino rats in 24hrs, 48 hrs, 72 hrs exposed to Snake

Naja naja venom

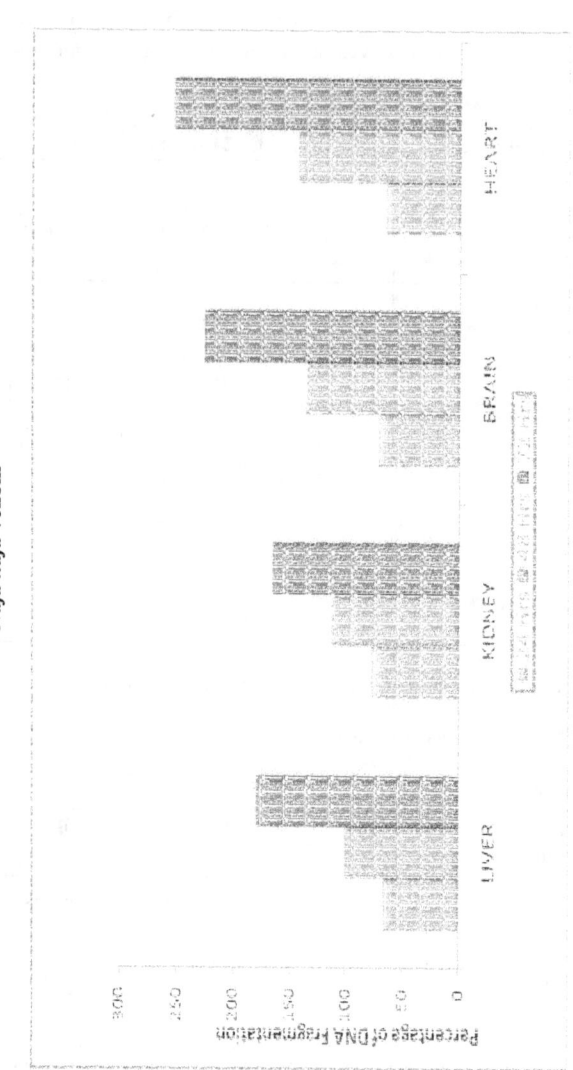

LEGEND FOR FIGURES

Plate. 3B :

Electrophoretic profiles of snake *Naja naja* venom treated albino rats in Liver, Kidney, Brain and Heart tissues.

M : Marker

C : Control

L : Liver

K : Kidney

B : Brain

H : Heart

L_1 to L_4 : 24 hrs of Snake *Naja naja* venom treated rats.

L_5 to L_8 : 48 hrs of Snake *Naja naja* venom treated rats.

L_9 to L_{12} : 72 hrs of Snake *Naja naja* venom treated rats.

Plate.3B: Electrophoretic profiles of Snake *Naja naja* venom treated Albino rats in Liver, Kidney, Brain and Heart tissues.

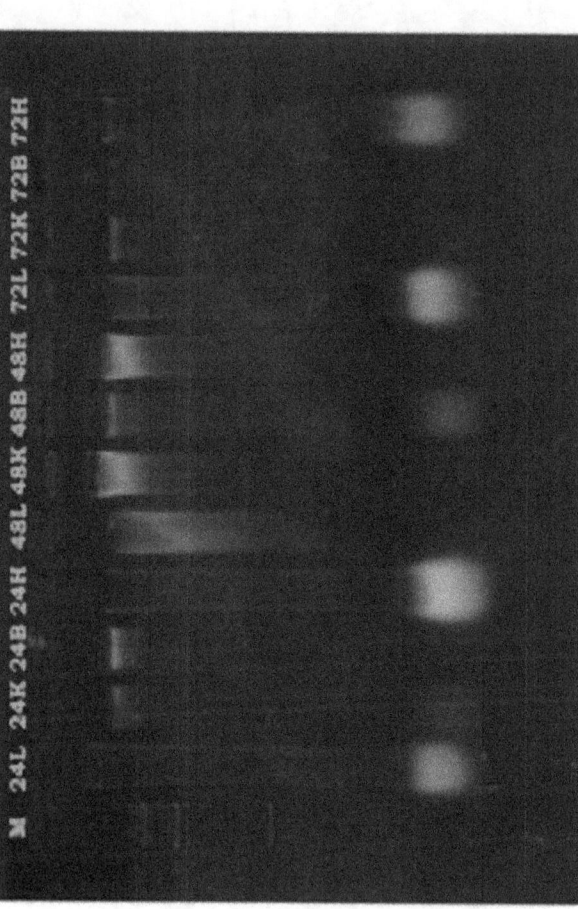

Chapter-IV

Detoxification Enzymes

Animal tissues are constantly coping with high reactive oxygen species, such as super oxide anion, hydroxyl radicals, hydrogen peroxides and other radicals generation during numerous peroxides during numerous metabolic reactions (Cabre *et al.,* 2000). The generation of small amount of free radicals appears to have an important biological function, but oxidative stress is caused by excess production of reactive species (Giardiano, 2005). To protect cell organ system of the body against reactive oxygen species, mammal cells are well equipped with a highly sophisticated and complex defense mechanism known both enzymatic and non enzymatic antioxidants.

Oxidative stress is defined as a disruption of the prooxidant, antioxidant balance in favor of the former, leading to potential damage. It is a result of one of the three factors: An increase in reactive oxygen species (ROS), an impairment of antioxidant defense systems and insufficient capacity to repair oxidative damage. Damage induced by ROS includes alterations of cellular macromolecules such as membrane lipids, DNA, and proteins. The damage may alter cell function through changes in intracellular calcium or intracellular pH, and eventually can lead to cell death (Kehrer *et al.,* 1990). Under normal conditions, excessive formation of free radicals and concomitant damage at cellular and tissue concentrations is controlled by cellular defense systems. These preventive defense systems can be accomplished by enzymatic and non enzymatic mechanisms including vitamin E and Glutathione. The antioxidant enzymes such as SOD, XOD and Catalase may also have an important function in mitigating the toxic effects of ROS (Adali *et al.,* 1999).

The first line of defense against O_2^- and H_2O_2 mediated injury are antioxidant enzymes; SOD, XOD and Catalase. The term antioxidant has been defined by Halliwell and Gutteridge (1990) as "any substance that delays or inhibits oxidative damage to a target molecule". Anti oxidant enzymes together with the substance that are capable of either reducing reactive oxygen metabolites (ROM) or preventing their formation, form a powerful reducing buffer which affects the ability of the cell to counteract the action of oxygen metabolites. All reducing agents there by form the protective mechanisms, which maintain the lowest possible levels of reactive oxygen metabolites inside the cell.

The molecular oxygen is used in many physiological processes, besides being the acceptor of electrons in the electron transport system. But at higher concentration, the oxygen is toxic to the body. The oxygen and its partially reduced forms can oxidize a variety of biomolecules in the body and exerts a sort of "oxidative stress". However, the body has a number of antioxidants to protect the cell from oxidative stress. Antioxidants can be defined as "compounds that protect biological systems against the potentially harmful effects of processes or reactions that can cause excessive oxidations".

Detoxification is a process of continuous reaction on particular chemical (Jacoby, 1980). Detoxification of xenobiotics (foreign antigens) includes two major steps. The primary phase involving oxidative, hydrolytic and other enzymatic pathways to produce polar end products. The secondary phase producing water soluble conjugates ready for excretion.

Free radicals are short lived chemical species produced in biological systems during antimicrobial defense through various oxidative enzymes such as Xanthine oxidase, monoamine oxygenase and by auto oxidation mediated by heavy metals and quinones (Batra *et al.,* 1989). These are atoms or molecules having odd number of electrons which make them highly reactive.These are highly destructive due to their reactive nature and ability to produce adducts or to cross link biological molecules. These free radicals are produced in the cells due to ionizing radiations and also during oxidative reactions constantly occurring in the cells. They perform a variety of functions namely oxidation of polyunsaturated fatty acids in cell membranes, lipid peroxidation, damage of DNA, deploymerization of hyaluronic acid and modulation of nucleotide cyclase activities.

Mammalian cells posses both enzymatic and non-enzymatic antioxidant defense mechanisms to cope up with oxygen free radicals. The enzymatic mechanism includes superoxide dismutase, catalase etc., (Venkataiah, 1995), where as non-enzymatic mechanism includes a variety of compounds as ascorbic acid and tocopherol etc., (Machlin and Bendich, 1987). When the production of reactive oxygen species exceeds the ability of the antioxidant system, it results in oxidative stress. To prevent cellular damage by free radicals, free radicals mediated lipid peroxidation and tissue antioxidants are essential.

Xanthine oxidase (XOD)

Xanthine is formed from the degradation of ATP and reoxygenation. Xanthine oxidase pathway is one of the important sites of free radical production. Xanthine oxidase generates oxygen radicals and uric acid from xanthine (MC Cord, 1993). Xanthine oxidoreductase under physiological conditions acts as a dehydrogenase (XDH). The dehydrogenase form of xanthine oxidoreductase under metabolic stress like hypoxia and ischemia converts to an oxidase form (XOD).

Xanthine oxidoreductase exists in two forms, as xanthine dehydrogenase (XDH) and Xanthine oxidase (XOD), which is formed through post translational modifications of XDH. Xanthine oxidase is unavailable to bind NAD and uses O_2 as its electron acceptor with both forms, but particularly with xanthine oxidase form, numerous ROS and RNS are synthesized (Vorbach *et al.,* 2003).

Xanthine oxidase was involved in major pathway of purine nucleotide catabolism in animals. It converts hypoxanthine and xanthine to uric acid. It is thought that xanthine oxidase has an important role in reperfusion injury. One of the mechanisms proposed is that under hypoxia conditions the depletion of cell ATP results in an elevated cytosolic concentration of AMP which is catabolized to adenosine, inosine and then hypoxanthine concomitantly the conversion of XDH in to XOD occurs by a protease, probably activated by an elevated cytosolic calcium concentration during ischemia. When, reperfusion takes place, the return of oxygen takes place to a production of Super oxide (Pasquier *et al.,* 1989).

Xanthine oxidase (XOD) is reported to play an important role in cellular oxidative stress, detoxification of aldehydes, oxidative injury in ischemia reperfusion, and neutrophils mediation. For example, XOD may serve as a messenger or mediator in the activation of neutrophil, T cell, cytokines or transcriptions in defense mechanisms rather than as after radical generator of tissue damage. Emerging evidence on the synergistic interactions of O_2^- a toxic product of XOD and nitric oxide.

Superoxide dismutase (SOD)

Super oxide dismutase (SOD) is the primary antioxidant enzyme in the cell and cellular defense against superoxide radicals. Among, other antioxidants enzymes, SOD considered as front line of defense against the potentially cytotoxic free radical oxidative stress. The SOD catalyzes the dismutation of two superoxide (O_2^-) radicals in to hydrogen peroxide (H_2O_2) and oxygen. These enzymes obey first order reaction kinetics and the forward rate constants are almost diffusion limited. This results in steady state concentration of Superoxide radicals in tissues that may vary directly with the rate of Super oxide generation and inversely with the tissue concentration of scavenging enzymes (Fattman *et al.*, 2003).

It is well known that SOD is involved in destroying the superoxide radicals and exists in several forms. Three isozymes of SOD exist in mammalian tissues. The enzyme substrate interaction in the mitochondria does not appear in auto inactivation mechanism (Karuzine and Archakov, 1994), indicating that mitochondrial SOD levels are maintained for longer period of time compared to the cytosolic form. Increased levels of SOD are generally taken as indirect evidence of an increased antioxidant milieu. Since, this is a sulfhydryl containing enzyme, decreased of its tissue level can also reflect oxidative denaturation. Due to its important role in detoxifying the superoxide anion radicals, this study was designed to find out to what extent SOD dismutates and the superoxide radicals undergo pesticide toxic stress.

SOD is the most important antioxidant enzyme because it is found virtually in all aerobic organisms. SOD's are a family of metalloenzymes that converts O_2^- to H_2O_2 according to the following reaction. The transition metal of the enzyme reacts with O_2^- taking its electron. O_2- is the only known substrate for SOD (Ray and Hussain, 2002).

$$O_2^- + O_2^- \xrightarrow[\text{SOD}]{2H} H_2O_2 + O_2$$

The super oxide dismutase (SOD) enzyme catalyzes the dismutation of two superoxide radicals into hydrogen peroxide and oxygen. The Hydrogen peroxide is further oxidized by enzymes. These enzymes obey first order reaction kinetics and the forward rate constants are almost diffusion limited. This results in a steady state concentration of super oxide in tissues that varies directly with the rate of superoxide generation and inversely with the tissue concentration of scavenging enzymes.

Catalase (CAT)

Catalase is one of the most important antioxidant enzyme, which can function either in the catabolism of hydrogen peroxide (H_2O_2) or in the peroxidative oxidation of substances, such as pesticides. Catalase has four sub-units. Each subunit contains a heme group; heme consists of a proto porphyrine ring and a central ferric atom (Fe^{3+}). The iron either in the ferrous (Fe^{2+}) or the ferric (Fe^{3+}) oxidative state and this heme group is responsible for carrying out catalase activity. To maintain catalytic activity CAT requires Fe^{2+} as a co-factor (Temel *et al.*, 2002).

Catalase is widely distributed enzyme in the body compartments, tissues and cells. In many cases the enzyme is located in sub cellular organelles such as, peroxisomes and cytosol (Leisuk *et al.*, 2003). Catalase is a tetrameric peroxidative enzyme which converts the hydrogen peroxide to water and molecular oxygen and whose gene expression is regulated by H_2O_2. Catalase plays an important role in ROS metabolism and an adaptation to oxidant stress (Vaziri *et al.*, 2003). Catalase catalyze the destruction of hydrogen peroxide into water and oxygen by the following reaction.

$$2H_2O_2 \longrightarrow 2H_2O + O_2$$

H_2O_2 is produced in the cells by a number of enzymatic reactions including those catalyzed by SOD, which converts superoxide anion radical to hydrogen peroxide and water.

Results

Xanthine oxidase (XOD)

The results of xanthine oxidase levels of the control and experimental albino rats under snake *Naja naja* envenomation are given in Table 4.1 and Figure 4.1. The experimental rats exposed to snake *Naja naja* venom showed statistically significant (P<0.01) increase of Xanthine oxidase levels in liver, kidney, brain, heart respectively. The increase in xanthine oxidase levels were time dependent manner in snake *Naja naja* venom treated rats.

In experimental conditions the tissues have shown increased xanthine oxidase levels in brain (86.25%) followed by liver (137.75%), heart (141.00%) and kidney (146.38%). The gradual increase was observed in 24 hrs, 48 hrs, 72 hrs of snake *Naja naja* venom treated rats. The lyotrophic series of increased XOD content in 72 hrs of snake *Naja naja* venom treated rats is as follows

Brain > Liver > Heart > Kidney

Superoxide dismutase (SOD)

The results of superoxide dismutase activity levels of the control and experimental albino rats under snake *Naja naja* envenomation are given in Table 4.2 and Figure 4.2. The experimental rats exposed to snake *Naja naja* venom showed statistically significant (P<0.01) decrease in superoxide dismutase activity levels in liver, kidney, brain, heart respectively. The decrease in superoxide dismutase activity levels were time dependent in snake *Naja naja* venom treated rats.

In experimental conditions the tissues have shown decreased superoxide dismutase activity levels in heart (58.03%) followed by kidney (47.07%), brain (31.72%) and liver (29.92%) in 72 hrs of snake *Naja naja* venom administered rats. The gradual decrease was observed in 24 hrs followed by 48 hrs and 72 hrs of snake *Naja naja* venom treated rats.

The lyotrophic series of superoxide dismutase activity decreased in 72 hrs of snake *Naja naja* venom treated rats is as follows

Heart > Kidney > Brain > Liver

Catalase (CAT)

The results of catalase activity levels of control and experimental albino rats under snake *Naja naja* envenomation are given in Table 4.3 and Figure 4.3. The experimental rats exposed to snake *Naja naja* venom showed statistically significant (P<0.01) decrease of catalase activity levels in liver, kidney, brain and heart respectively. The decrease in catalase activity levels were time dependent in snake *Naja naja* venom treated rats.

In experimental conditions the tissues have shown decreased catalase activity levels in liver (65.21%) followed by kidney (52.71%), heart (40.99%) and brain (36.661%) in 72 hrs of snake *Naja naja* venom administration. The gradual decrease was observed in 24 hrs followed by 48 hrs and 72 hrs of snake *Naja naja* venom treated rats.

The lyotrophic series of catalase activity decrement in 72 hrs of snake *Naja naja* venom treated rats is as follows

Liver > Kidney > Heart > Brain

Discussion

Xanthine Oxidase (XOD)

    In the present investigation the xanthine oxidase levels were increased in all the tissues of rat in 24 hrs, 48 hrs and 72 hrs of envenomated rats. Under snake *Naja naja* venom stress, significant increased xanthine oxidase activity (Table 4.1) might be due to conversion of xanthine dehydrogenase to xanthine oxidase. For nitrogen balance of the tissue, xanthine oxidase is produced when the native form of xanthine dehydrogenase is altered either by sulphydryl oxidation or by limited proteolysis (Dellacorte and Stripe, 1972). During the apoptosis in rat mammary gland, the mitochondrial XOD activity was increased (Rus *et al.*, 2007).

    The elevated levels of xathine oxidase in the present investigation indicate the over production of superoxide anions (O_2^-) in the different tissues of albino rat in response to snake *Naja naja* venom envenomation. In *Boleophthalmus pectinirostris* liver, the heavy metal cadmium (Cd^{2+}) caused an increased XOD activity levels (Liu *et al.*, 2006).

    XO is a ubiquitous enzyme existing in most normal functioning tissues, mainly in the XDH form. However, once the tissue is exposed to a metabolic stress, such as inflammation, hypoxia or ischemia, the enzyme converts to the oxidized form. In the presence of adequate amounts of the substrate and oxygen, XO enables the generation of cytotoxic oxidants such as superoxide anion($O2^-$), hydrogen peroxide (H_2O_2) and hydroxyl radical (OH). The entire cascade of XO generated oxidative events was proven to be damaging in animals. The elevated total XO activity in the liver, kidney, brain, heart were documented in the current study is suggestive of stress oxidant mediation of multiple organ dysfunction that followed envenomation. The absence of histological damage in our study is probably due to the short time that elapsed between the injection of the venom and the effect of venom to the target organs. These findings, therefore, do not exclude this cascade of the events.

The XO capability and the concentration of XO on the cell surfaces might become a mechanism that would supply the circulation with oxidants uninterruptedly. This concept of binding and concentrating XO within tissues and later propagating the damage. Snake *Naja naja* venom in the rat, results in systemic circulatory deterioration and the effects of multi-organ damage. TNF and XO were involved in damaging processes. Clarification of their combined effect may pave the way for new and enhanced therapeutic modalities for patients with snake bites.

Superoxide dismutase (SOD) and catalase (CAT)

Superoxide dismutase (SOD) and catalase (CAT) are involved in the detoxification of reactive oxygen species generated during the snake *Naja naja* venom administration, the SOD and catalase activities were estimated in different tissues of albino rat. Both the enzyme activities were decreased with increase of time schedules. The decreasing activity reflects the oxidative status of different tissues. Because of the high concentration of Poly Unsaturated Fatty Acids (PUFA) and aerobic metabolic activity of different tissues, increase the susceptibility of these organs to peroxidative damage induced by reactive oxygen species after snake *Naja naja* venom administration. So due to snake *Naja naja* venom impact on different tissues (liver, kidney, brain, heart) undergo damage through free radicals. This oxidative stress produces depleted activity of both the antioxidant enzymes such as superoxide dismutase and catalase.

Superoxide dismutase and catalase are generally involved in the detoxification of superoxide anion radical generated by xanthine oxidase. In the present study the superoxide dismutase activity was decreased according to time schedules. This result was in agreement with the result of Manna *et al.,* (2005). The superoxide dismutase and catalase levels were decreased significantly in different tissues.

The basis of Snake *Naja naja* venom toxicity in the production of reactive oxygen species may be due to their Redox–cycling activity, they readily accept an electron to form free radicals and then transfer them to oxygen to generate Superoxide anions and hence H_2O_2 formation through dismutation reaction. Generation of free radicals probably because of the alterations in the normal homeostasis of the body resulting in oxidative stress, if the requirement of continuous antioxidants is not maintained (Ryrfeldt *et al.*, 1992). Hexachloro-hexane (HCH) effect on immature chick tissues decreased SOD activity. A gradual decrease in catalase activity was observed after Isoproterenol administration in to the tissues of rats (Rathore *et al.*, 2000). Catalase activity decreased significantly in the cyfluthrin treated tissues of albino rats. This means hydrogen peroxide generation plays important role in the toxicity of cyfluthrin (Omotuyi, 2006). Effects of some environmental parameters on catalase activity measured in the mussel (*Mytilus galloprovincialis*) exposed to lindane (Khessiba *et al.*, 2005).

Ferrari, (2007) reported the decreased catalase content in liver and kidney of rainbow trout. The early inhibitory effect in CAT activity may be associated with a high degree of oxidative stress. According to Alka Gupta *et al.*, (1999) in albino rat on exposure to quinolphos, SOD and catalase activities were decreased as 63% and 31% respectively. A synthetic, pyrethroid deltamethrin significantly decreased the SOD and catalase activities in albino rats (Manna *et al.*, 2005).

Thus the data provide evidences for induction of oxidative stress on Snake *Naja naja* venom exposure. Xanthine oxidase (XOD) activity levels were elevated in all the tissues with increase on time schedules which was due to the over production of oxygen free radicals. Superoxide dismutase (SOD) activity was decreased in all the tissues with increase on time schedules. The SOD activity reflects the oxidative status of tissues, there by the cells of different tissues were damaged by free radicals. This oxidative stress produces depletion of SOD activity. Catalase activity (CAT) was decreased with an increase on time schedules which was due to cell damage resulting in oxidative stress. It is due to cell damage resulting in oxidative stress leading to the depletion of catalase activity.

Table 4.1:   **Changes in Xanthine Oxidase (XOD) activity (μ moles of formazon formed/mg protein/hr) in different tissues of control and snake *Naja naja* venom treated albino rats. Values in parentheses indicate percent change over control.**

Name of the tissue	Control	24 Hours	48 Hours	72 Hours
Liver				
Mean	0.833	1.167	1.180	1.980
SD	±0.359	±0.010	±0.014	±0.013
PC		(40.192)	(41.653)	(137.750)
Kidney				
Mean	0.649	1.001	1.307	1.598
SD	±0.013	±0.011	±0.013	±0.012
PC		(54.408)	(101.516)	(146.389)
Brain				
Mean	0.412	0.470	0.538	0.768
SD	±0.009	±0.013	±0.011	±0.013
PC		(13.905)	(30.356)	(86.257)
Heart				
Mean	0.311	0.599	0.693	0.748
SD	±0.013	±0.010	±0.008	±0.013
PC		(92.915)	(123.081)	(141.009)

All the values are mean ± SD of six individual observations.
SD – Standard Deviation.
PC – Percent change over control.

ONE WAY ANOVA

Source of Variation	DF	Liver	Kidney	Brain	Heart
		MS	MS	MS	MS
Between Groups	3	4.272**	2.989**	0.438**	0.683**
Within Groups	20	0.647	0.003	0.003	0.003
Total	23				

NS: Not Significant, *-Significant ($P<0.05$), **- Highly Significant ($P<0.01$)

Fig. 4.1 : Xanthine oxidase activity levels in different tissues of albino rats exposed to snake *Naja naja* venom.

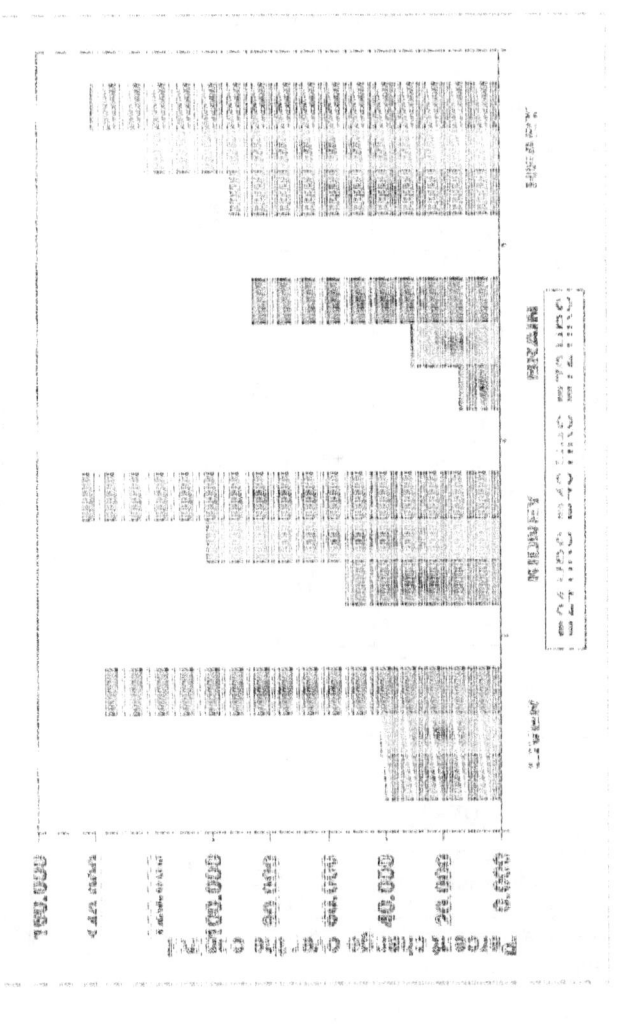

Table 4.2:   Changes in Superoxide Dismutase (SOD) activity (units of superoxide anion reduced/mg protein/min.) levels in different tissues of control and snake *Naja naja* venom treated albino rats. Values in parentheses indicate percent change over control.

Name of the tissue	Control	24 Hours	48 Hours	72 Hours
Liver				
Mean	7.112	6.171	5.255	4.983
SD	±0.008	±0.011	±0.014	±0.012
PC		(-13.220)	(-26.113)	(-29.928)
Kidney				
Mean	7.797	6.149	5.150	4.127
SD	±0.018	±0.014	±0.019	±0.033
PC		(-21.127)	(-33.951)	(-47.070)
Brain				
Mean	2.123	2.098	1.750	1.449
SD	±0.010	±0.013	±0.012	±0.009
PC		(-4.186)	(-17.580)	(-31.729)
Heart				
Mean	2.259	1.965	1.001	0.948
SD	±0.358	±0.014	±0.019	±0.019
PC		(-12.995)	(-55.671)	(-58.033)

All the values are mean ± SD of six individual observations.
SD – Standard Deviation.
PC – Percent change over control.

ONE WAY ANOVA

Source of Variation	DF	Liver	Kidney	Brain	Heart
		MS	MS	MS	MS
Between Groups	3	16.782**	43.987**	1.838**	8.027**
Within Groups	20	0.003	0.010	0.003	0.646
Total	23				

NS: Not Significant, *-Significant ($P<0.05$), **- Highly Significant ($P<0.01$)

Fig. 4.2 : Superoxide dismutase activity levels in different tissues of albino rats exposed to snake *Naja naja* venom.

Table 4.3: Changes in Catalase activity (μ moles of H_2O_2 decomposed /mg protein/min) levels in different tissues of control and snake *Naja naja* venom treated albino rats. Values in parentheses indicate percent change over control.

Name of the tissue	Control	24 Hours	48 Hours	72 Hours
Liver				
Mean	0.399	0.305	0.207	0.139
SD	±0.015	±0.007	±0.007	±0.013
PC		(-23.495)	(-48.077)	(-65.217)
Kidney				
Mean	0.267	0.210	0.165	0.126
SD	±0.015	±0.010	±0.011	±0.014
PC		(-21.585)	(-38.116)	(-52.714)
Brain				
Mean	0.189	0.148	0.130	0.120
SD	±0.015	±0.013	±0.010	±0.011
PC		(-21.731)	(-31.360)	(-36.661)
Heart				
Mean	0.187	0.175	0.157	0.110
SD	±0.015	±0.010	±0.014	±0.011
PC		(-6.595)	(-16.310)	(-40.998)

All the values are mean ± SD of six individual observations.
SD – Standard Deviation.
PC – Percent change over control.

ONE WAY ANOVA

Source of Variation	DF	Liver	Kidney	Brain	Heart
		MS	MS	MS	MS
Between Groups	3	0.233**	0.066**	0.017**	0.020**
Within Groups	20	0.003	0.003	0.003	0.003
Total	23				

NS: Not Significant, *-Significant (P<0.05), **- Highly Significant (P<0.01)

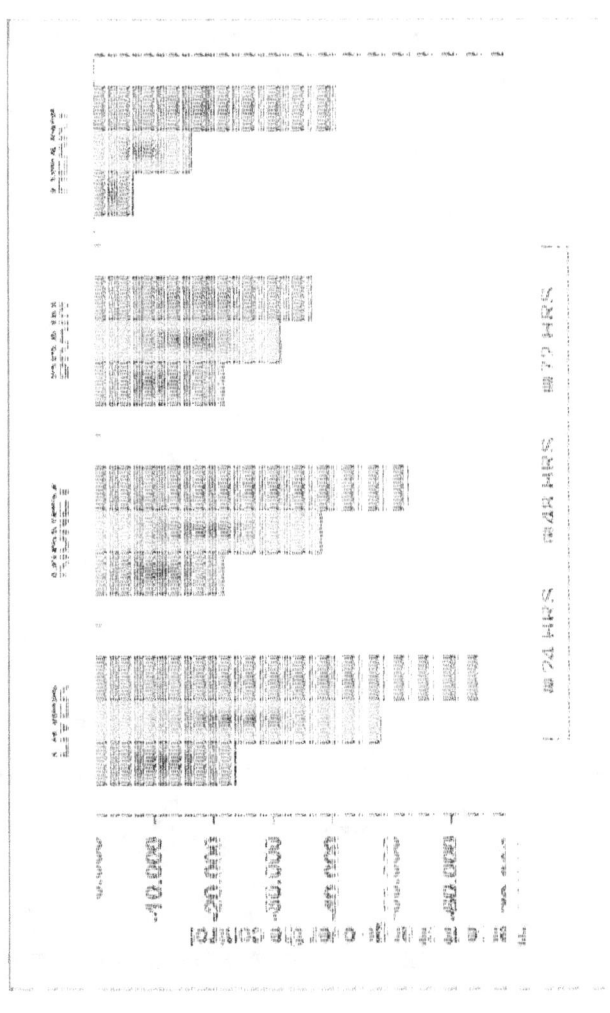

Fig. 4.3 : Catalase activity levels in different tissues of albino rats exposed to snake *Naja naja* venom.

Chapter-V

Histopathology
Light microscopy

Histology, is the study of micro anatomy of specific tissues, has been successfully employed as a diagnostic tool in medical and veterinary science, first cellular investigation were carried out in the mid nineteenth century. Exposure of animals to contaminated water also causes severe pathological changes at the tissues level. Snake venom enter the body via internal digestive system after oral administration. Venoms are not subjected initially either to the detoxifying reactions of the liver or to excrete via the biliary system. Compounds transported by oral feeding effect can be distributed to all parts of the body in their unmetabolised form (Turner and Shanks, 1980).

The examination and study of normal cells and tissues by microscopy is called histology. The study of abnormal cells and tissues is known as histopathology (Aughey and Frye, 2001). Toxicological histopathology gives useful data concerning the changes induced by chemicals at the tissue and cellular level. All the tissues and organs in the body of an animal may be potential targets for the toxic effects of any chemical or metal. A histo-pathological assessment throws light on the nature of tissue alteration and the extent of damage. This in turn indicates the toxic nature of the compound. Therefore, histology gives useful insight into the tissue lesions prove to the external manifestations of the deleterious effects of heavy metals.

Snake venom possess high toxicity not only to target organisms but also to non-target organisms. These substances find their way in to places far from application and lead to alterations in metabolic activities of living organisms by bio-accumulation. Commonly venom accumulate to a greater extent in the liver (Edwards, 1973) which is the centre for snake venom metabolism. Venomous residues in the tissues cause serious physiological alterations even at low levels.

Venoms which are ubiquitous in nature have become integral part in the tissues of animals. Venom find their way in the application and accumulate in significant concentrations in the tissues of animals. The extent of severity of tissue damage is a function of the concentration and potentiality of toxic

140

compound accumulated in the tissues is a time dependent. It is obvious that any venom indult could cause pathological damage or injury to cells in an animal if it is taken beyond the safe permissible limits. Susceptibility to venom injury varies greatly among the tissues and cells of the same animal and also varies in different animal groups.

The physiological investigations may not help in the complete understanding of the venom impact on a tissue. When coupled with cytoarchit-ectural studies, the toxicological studies will give a clear understanding of the venom impact on the tissues.

Tissue changes following snake envenomation depend on the species of snake responsible for the bite, the composition of its venom and also the susceptibility of the tissue for a particular component of the venom (Kamiguti *et al.*, 2000). Histologically, changes due to viper venom have been seen in all renal structures (Soe *et al.*, 1993). Degeneration, necrosis and regeneration of tubular epithelial cells are seen in renal tubules. Interstitial oedema, cellular infiltration, arthritis, thrombophlebitis, congestion, infiltration and cortical necrosis have also been observed (Sitprija, 2006). In the liver, congestion and petechial haemorr-hages, microvescicular fatty change, hydropic degeneration as well as necrosis of hepatocytes have been recorded following snake bites (De Silva *et al.*, 1992).

Myocardial changes are common when snake venom is rich in cardiotoxins. Congestion of myocardial blood vessels and petichial haemorr-hages are the frequent histological manifestations. Skeletal muscles are affected in the presence of myotoxic substances in the venom. Microscopically, congestion of blood vessels and necrosis are the common findings. Indirectly, muscle damage results in myoglobinuria which could be detected biochemically.

Histological changes in the brain tissues following snake bites are not widely documented compared to other organs. However, since snake venom is rich in neurotoxins significant microscopical changes are likely to be present in the brain tissue. Multiple small foci of haemorrhages and congestion of blood vessels have been reported following administration of Collubrid snake venom to rats which was more marked with intravenous injection of venom (Peichoto *et al.*, 2006). Apoptosis of Schwann cells following intramuscular injection of beta bungarotoxin has been demonstrated in chick embryos (Ciutat *et al.*, 1996). In addition, histopathological changes have been demonstrated in the lungs and adrenal glands (Kularatne and Ratnatunga, 2001). *Bungarus* venom is a rich source of phospholipaseswhich break down phospholipids in the cell membrane. This is thought to be the most important mode of tissue damage in elapid snake bites (Sitprija, 2006).

The objective of this study was to identify and compare the histological changes in liver, kidney, brain and heart tissues following the oral intubation of the snake *Naja naja* venom.

In the present study, 48 hrs histopathological studies were commonly observed in 24 hrs and 72 hrs studies, hence 48 hrs microphotographs were not described here and focused on 24 hrs and 72 hrs only.

RESULTS

Normal histology of rat liver

Liver is partially divided in to three hepatic lobes and incompletely covered by tunica serosa and a delicate lobe has two main constituents, an epithelial parenchyma and system of blood sinusoids. Liver contain large number of hexagonal functional units called lobules. The classic lobule is traditionally described as roughly cylindrical with a venous channel, the central that course through its long axis. Irregular inter connecting sheets, a plate like arrangement of hepatic cells or hepatocytes radiate outward from the central vein and constitute the parenchyma of the lobule. Sinusoids separate the sheets of hepatic cells and empty in the central veins. At the angles of the hexagons are the portal canals which are loose stromal connective tissue characterized by the presence of the portal triads. It also has the lymphatic vessel. Connective tissue of portal area is ultimately continuous with the fibrous capsule of liver (Plate. 5.1, Fig.A). The portal canal is bordered by the outermost hepatocytes of the lobule. Sinusoids are lined by endothelial lining. Endothelium of sinusoids is discontinuous due to presence of large fenestrae and large gaps lining of sinusoids also contain second type of cells, called stellate sinusoidal macrophages of kupffer cells.

Histopathological changes of rat liver under snake *Naja naja* venom intoxification

The microscopic examinations revealed that snake *Naja naja* venom induced histopathological lesions in liver, after 24 hrs and 72 hrs, of venom administration. The degree of severity differed from 24 hrs to 72 hrs of envenomation of snake *Naja naja* venom and it was more in the later period than former.

In 24 hrs snake *Naja naja* venom administration in liver, central vein congestion, dilated sinusoids were observed (Plate. 5.1, Fig. B) and dilated and engorged hepatic portal vein was observed (Plate 5.1; Fig. C), and hepatic cell, plasma membrane, nucleus of the hepatocyte cell were observed (Plate 5.1, Fig. D) .

In 72 hrs of snake *Naja naja* venom administration, diffused necrotic areas and severe degenerative changes in central vein, inflammatory foci were observed (Plate 5.1, Fig. E) and hepatic portal vein with thrombosis (star shaped) (HPVT) were observed (Plate 5.1, Fig. F) and sinusoidal haemorrhage (SH), focal necrotic areas (FNA), amyloid precipitation (AP), hepatocyte cell (HC) with prominent nucleus were observed (Plate.5.1, Fig.G).

Normal histology of rat kidney

Each kidney is enclosed in a tough connective tissue capsule extending into the parenchyma and has two regions the cortex and the medulla.

The nephron is the functional unit of the kidney. The major subdivisions are the renal corpuscle and the uriniferous tubule.

The blind end of the proximal tubule is identified with a network of capillaries and supporting cells to form a filtering system, the renal corpuscle. Each renal corpuscle consists of a glomerulus and a glomerular (Bowman's) capsule. The outer layer of the glomerular capsule is the capsular (parietal) wall, which is separated from the glomerular (visceral) layer by the capsular (urinary) space. The capillaries of the glomerulus are served by an afferent and an efferent arteriole, entering and leaving the renal corpuscle at the vascular pole. At the opposite pole is the capsular space, where the filtrate passes into the proximal tubule at the urinary pole of the renal corpuscle.

This nomenclature is based on the functional areas of the renal tubule. The proximal convoluted tubule (PCT) is long and lined with low columnar cells with a basal nucleus. The cytoplasm is deeply stained with eosin and the apical surface is a continuous brush border. The PCT is continued with the proximal straight tubule. It is similar in appearance and extends towards the medulla where the epithelium changes abruptly to simple squamous. This part of the tubule descends into the medulla as the thin descending limb and bends sharply to return to the cortex as the thick ascending limb, which was previously known as the "loop of Henle." In the cortex the epithelium becomes cuboidal or columnar and forms the distal straight tubule and coils near the glomerulus to become the distal convoluted tubule (DCT). The DCT is shorter than the PCT, the epithelium is cuboidal, the cytoplasm is paler and there is no brush border.

The DCT approaches the glomerulus at the vascular pole, where it thickens, and the cell nuclei of the tubule wall become crowded together to form the macula densa, part of the juxtaglomerular apparatus. Juxta glomerular cells are modified smooth muscle cells in the walls of afferent arterioles close to the glomerulus. The collecting tubule or duct (lined with poorly staining cuboidal epithelium is the terminal segment of the nephron, a continuation of the OCT within the medulla (Plate 5.2, Fig. A).

Histopathological changes of rat kidney under snake *Naja naja* venom intoxification

The microscopic examinations revealed that snake *Naja naja* venom induced histopathological lesions in kidney, after 24 hrs, and 72 hrs of snake *Naja naja* venom administration and the degree of severity differed from 24 hrs to 72 hrs of snake *Naja naja* envenomation.

In 24 hrs of snake *Naja naja* venom administration, slight thickening of bowman's capsule, infiltration, glomeruli stalk, glomeruli congestion of blood vessel, hypertrophy of bowman's capsule, fragmentation and atrophy of glomerulus, were observed (Plate 5.2, Fig. B-D). In 72 hrs of snake *Naja naja* venom administration, fragmentation of glomerulus and increased size of the lumen, hypertrophy of bowman's capsule, congestion of blood vessel were observed (Plate 5.2, Fig. E to F) .

Normal histology of rat cerebral cortex

The cerebral cortex is the largest part of the vertebrate brain and is the source of neural transmissions that enhance memory, plasticity, cognition, speech and intellectual activity. The cytoarchitectural structure of cortex is characterized by the presence of six – layered laminated pattern of cells. 1st layer – consists of mostly glial cells, axons of neurons of other layers and very few neurons. 2nd layer – consists of small pyramidal cells. 3rd layer - consists of large pyramidal cells. 4th layer – consists of stellate and granule cells which receive input to the cortex from thalamo cortical fibers, association fibers and commissural fibers. 5th layer – consists of largest pyramidal cells known as giant pyramidal cells or betz cells. 6th layer–consists of martinotti cells.

Microphotograph of control rat cerebral cortex shows different layers such as molecular layer (ML) with glial cells, outer pyramidal layer (OPL) with large and small pyramidal cells and outer granular layer (OGL) with stellate and granular cells. Below the granular layer, betz cells are present which are considered as the largest pyramidal cells. (Plate 5.3, Fig. A).

Histology of rat cerebral cortex under Snake *Naja naja* venom intoxification

The cerebral cortex under 24 hrs of Snake *Naja naja* venom administration shows mild pycnosis of the cytoplasm of the neurons and mild congestion in neurofibrillar network. These deformation levels were increased in 72 hrs cerebral cortex (Plate.5.3, Fig.C). In 72 hrs, the observed deformations are severe pycnosis of the neuron cytoplasm, severe congestion in blood vessels, degenerated glial cells, vacuoles are formed due to the degeneration of the mitochondria and golgi apparatus of glial cells. Congestion of the neurofibrillar network (CNFN), the congested blood vessels (BVC), are due to severe pycnosis (Plate. 5.3, Fig.D).

Normal histology of rat cerebellum

One of the most impressive parts of the brain is cerebellum, located at the lower back of the brain and is more rapidly acting mechanism than any part of the brain. The cerebellum is not only involved in skilled motor performances but also involved in various sensory functions including sensory acquisition, discrimination, tracking, prediction etc.

The three functional regions of cerebellum are vestibulo- cerebellum, spino cerebellum and cerebro cerebellum. The cerebellum can be divided into three cortical layers with the same basic neuronal circuitry every where which involve five main cell types as follows.

1) Outer molecular layer – Basket cells and Stellate cells

2) Middle purkinje layer – Purkinje cells (Largest neurons)

3) Inner granule layer– Granule cells and Golgi cells

Microphotograph of control rat cerebellum shows outer molecular layer (OML), middle purkinje cells layer (MPCL) and inner granule cell layer (IGCL). Basket and Stellate cells are present in OML and Purkinje neurons are present in MPCL. (Plate. 5.3, Fig. E).

Histology of rat cerebellum under Snake *Naja naja* venom intoxification

The cerebellum under 24 hrs venom administration (Plate. 5.3, Fig. F) shows slight degenerative changes in granular layer and in purkinje cells. In s72 hrs venom administration, severe degeneration of purkinje cells and severe necrotic changes in granular layer (NGL) and molecular layer were observed (Plate.5.3, Fig.G).

Normal histology of rat hippocampus

The hippocampus is a brain structure which lies under the medial temporal lobe one on each side of brain and is part of limbic system. Hippocampus plays significant role in the formation of long-term memories. Hippocampus is grouped with nearby structures including dentate gyrus and is called hippocampal formation. The hippocampal formation is bilateral structure sandwiched between the cortex and the thalamus.

Hippocampal formation consists of hippocampus proper, the dentate gyrus, the subicular complex and the fornix. Hippocampus proper (Cornu Ammonis) is subdivided in to four main cytoarchitectural fields namely CA_1, CA_2, CA_3 and CA_4 that are unidirectionally connected from CA_4 to CA_1. CA_2 and CA_4 are small and not well defined.

The neurons of hippocampus have spatial firing fields called Pyramidal cells. Pyramidal cells are present in CA_3 and CA_1 regions.

Some important anatomical features of hippocampus are as follows.

147

1) Dentate gyrus possesses 1.2 million granule cells, 4K basket cells, 32 K hilar interneurons and 20 K mossy cells.

2) CA_3 subfield has16 x 10^3 Pyramidal cells and CA_1 consists of 250 x 10^3 Pyramidal cells. Together CA_3 / CA_1 have 330K / 420K pyramidal cells and various interneurons.

3) Subiculum possesses around 180 K cells.

Hippocampal formation is formed of Enthorhinal cortex, dentate gyrus, CA_3, CA_1 and Subiculum.

Microphotograph of control hippocampus shows characteristic curvature of hippocampus. Cornu Ammonis layers (CA_1 and CA_3) are separated by compact glial cells. Between the nerve cells a thick neurofibrillar net work (NFN) known a neuropile (NP) is present. Pyramidal cells are present in CA_3 and CA_1 region. (Plate. 5.3, Fig. H).

Histology of rat hippocampus under Snake *Naja naja* venom intoxification

The hippocampus under 24 hrs of Snake *Naja naja* venom administration (Plate. 5.3, Fig. I) shows slight pycnosis (P), congestion in neurofibrillar network (CNFN) and blood vessel congestion (BVC). These deformations are severely increased from 24 hrs of venom administration to 72 hrs of venom treated hippocampus. In 72 hrs envenomated rats, (Plate. 5.3, Fig. J) the observed deformations were pycnosis (SP), congested glial cells (CGC), neurotic degeneration in CA3 layer (NC) and blood vessel congestion (BVC) in CA1 layer.

Normal histology of rat heart

The cardiac wall consists of three layers: endocardium (inner), myocardium (middle) and epicardium (outer). The endocardium contains continuous squamous endothelial cells, vascular areolar connective tissue and conducting fibres. The myocardium is composed of cardiac muscle and also

148

contains vascular areolar connective tissue. The epicardium is thicker than the endocardium, and fat deposits in the dense connective tissue and coronary blood vessels are often found. Fibrous rings support the heart valves. They provide a means of insertion for the cardiac muscle and may be referred to as the fibrous or cardiac skeleton (Plate. 5.4, Fig. A).

Histopathological changes of rat heart under snake *Naja naja* venom intoxification

The microscopic examinations revealed that snake *Naja naja* venom induced histopathological changes in rat heart, after 24 hrs, 72 hrs of snake *Naja naja* venom administration, the degree of severity differed from 24 hrs to 72 hrs. In 24 hrs of snake *Naja naja* venom administered rats, the heart did not show any marked pathological changes (Plate 5.4, Fig.B). Under 72 hrs of snake *Naja naja* venom administration, diffused area with hemorrhage and rounded nucleus were observed (Plate 5.4, Fig. D to F).

DISCUSSION

Histopathological changes were seen in the liver, kidney, brain, and heart and those were of an acute toxic insult. Snake *Naja naja* venom is rich in neurotoxins and certain neurotoxins like kappa-toxin have its primary action in the central nervous system. This would be contributory to the presence of histopathological changes in the brain.

Post synaptic toxins get concentrated in the kidney at higher concentrations than in other tissues (Chang *et al.,* 1979). On the other hand, the main route of excretion of Snake *Naja naja* venom toxins is *via* kidney. This would explain the presence of congestion and inflammation in the kidneys. However, absence of nephrotoxic agents, vasculotoxins and haemotoxins would be the reason for the lack of necrosis in the kidneys. Liver, on the other hand being the primary detoxifying organ in the body could be affected by many types of toxic components in venom. However, since the affinity of neurotoxins is lower in the liver, it is possible that these changes are resulting from another toxic component in the venom other than neurotoxins. Snake *Naja naja* venom also contains phospholipases which show a range of different actions on different tissues. These enzymes hydrolyze phospholipids in the cell membrane and disturbs the cell membrane activity (Iwanaga and Suzuki, 1979). In the present study, the histopathological manifestations, mainly inflammatory and necrotic in nature could have occurred in response to the membrane damage caused by the action of these phospholipases. Snake *Naja naja* venom, which is rich in phospholipases causes extensive tissue damage involving many organs.

Lower molecular weight toxins in snake *Naja naja* venom are rapidly absorbed and fastly distributed through the blood stream. Thus, toxic changes would appear in the target organs clearly. In this study, the tissue changes were evident even in 24 hrs hour after venom injection in the liver, kidney, brain, as well as the heart.

Hyaluronidase, is known to enhance tissue distribution of other toxins (Iwanaga and Suzuki, 1979). Although snake *Naja naja* venom is not a rich source of hyaluronidase, *B. ceylonicus* and *B. caeruleus* also would contain hyaluronidase in their venom leading to a faster action. This could be the reason particularly with snake *Naja naja* venom which has led to prominent tissue changes even at much lower doses.

Rats injected with snake *Naja naja* venom, inflammatory infiltration, necrosis were observed in liver tissues. In the kidney, inflammatory changes were found to increase with concentration. Administration of snake *Naja naja* venom resulted in congestion, inflammation and necrosis in the brain in all experimental groups.

Since crude venom was injected, differences in venom composition could have contributed to the above differences. Necrosis of liver tissue produced by snake *Naja naja* venom suggests the possibility of a cardiotoxin like substance. This is supported by the fact that cardiotoxin in cobra venom has a higher affinity towards the liver than the brain and kidney and heart and it was supported by Chang (2003).

A previous study in India using *B. caeruleus* venom, has demonstrated haemorrhage and necrosis of proximal tubules of the kidney and petechial haemorrhages in the myocardium (Kiran *et al.*, 2004) in addition to findings similar to the present study.

The studies on the effect of Snake *Naja naja* venom in heart tissue is scanty, however effect of toxins (i.e., pesticides) are abundant, Savithri (2009) studied the toxic effect of chlorpyrifos in albino rat observed the congestion and slight infiltration in cardiac muscle exposed to chlorpyrifos. The literature available on impact of snake venom on brain is hardly available, however Siraj (2008) studied the influence of pesticide acephate in different brain regions of albino rat and observed slight vacuolation, mild degenerative changes in granule cells, vacuolative changes with mild congestion of blood vessels in molecular layer, slight degenerative changes in neurofibrillar network etc., were observed in cortex, cerebellum, hippocampus. Similar changes were observed in the present study.

The present study suggest that the snake *Naja naja* venom induces moderate histopathological changes in vital organs of rat such as liver, kidney, brain, and heart after its oral intubation. These changes are intiated at early stages of the envenomation and may be associated with a behavioral or functional abnormality of those organs during envenomation. Moreover, these damages may lead to permanent sequelae. As considerable caution should be excercised in extrapolating experimental studies in animals to human enve-nomation, it would be interesting to determine whether snake *Naja naja* venom acts similarly in Human victims.

151

LEGEND FOR FIGURES

Plate 5.1

Fig. A: Control rat liver showing Hepatocytes (H) with centrally placed prominent Nucleus (N) with Sinusoids (S) and Central Vein (CV). H & E. 100 X.

Fig. B: 24 hours of Snake *Naja naja* venom administered rat liver showing Central Vein Congestion (CVC), Dilated Sinusoids (DS). H & E. 100 X.

PLATE – 5.1

Fig.A

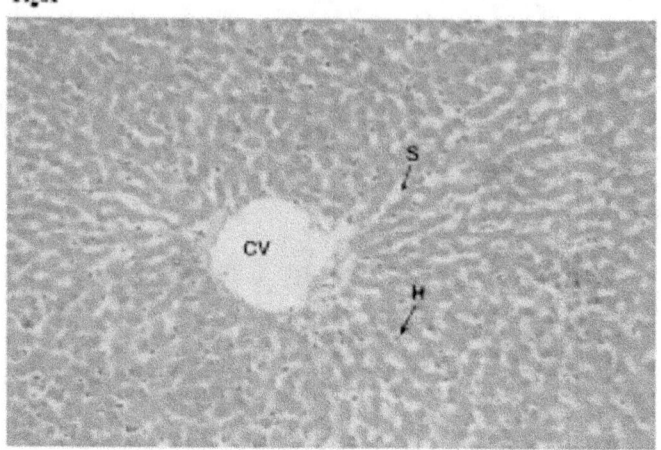

Fig. B

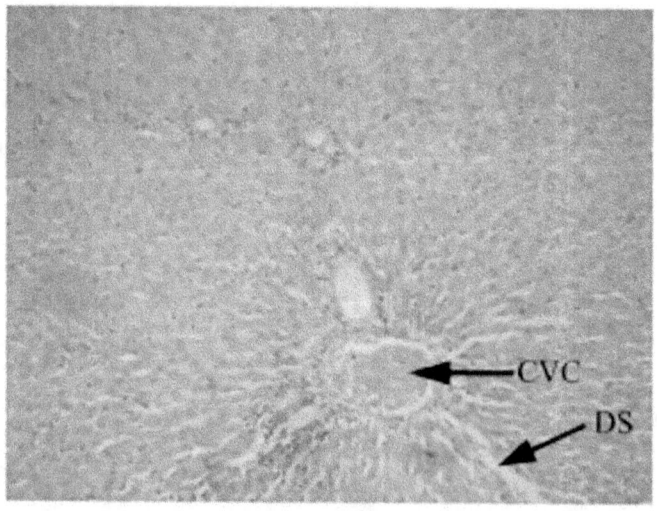

LEGEND FOR FIGURES

Plate 5.1

Fig. C: In higher magnification 24 hrs Snake *Naja naja* venom administrated rat liver showing, Central Vein Congestion (CVC), and Dilated Sinusoids (DS), Dilated and Engorged Hepatic Portal Vein (DEHPV). H & E. 400 X

Fig. D: 24 hrs Snake *Naja naja* venom administrated rat liver showing Hepatocyte cell (HC), Plasma Membrane (PM), Nucleus of the Hepatocyte Cell (N), Lobular Disarray (LDA). H & E. 400 X.

PLATE - 5.1

Fig. C

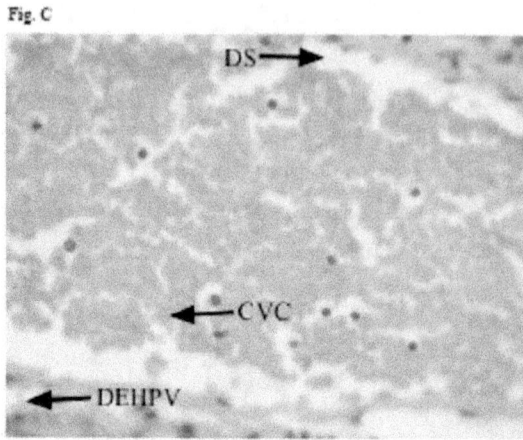

Fig. D

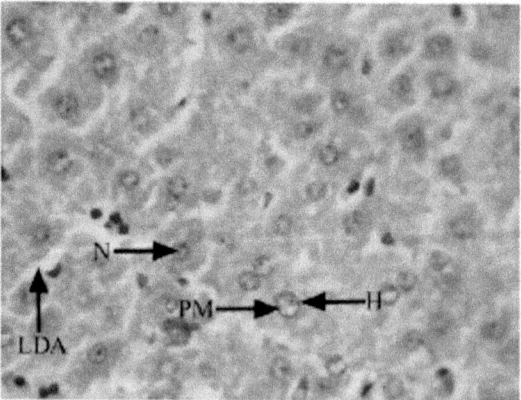

LEGEND FOR FIGURES

Plate 5.1

Fig. E: 72 hrs of Snake *Naja naja* venom administrated rat liver showing Diffused Necrotic Areas (DNA), and Severe Degenerative Changes in Central Vein (SDCV), Inflammatory foci (IF). H & E. 100 X.

Fig. F: In higher magnification of 72 hrs of Snake *Naja naja* venom administrated rat liver showing Diffused Necrotic Areas (DNA) and Severe Degenerative Changes in Central Vein (SDCV), Hepatic Portal Vein With Thrombosis (Star shaped) (HPVT). H & E. 400 X.

PLATE – 5.1

Fig. E

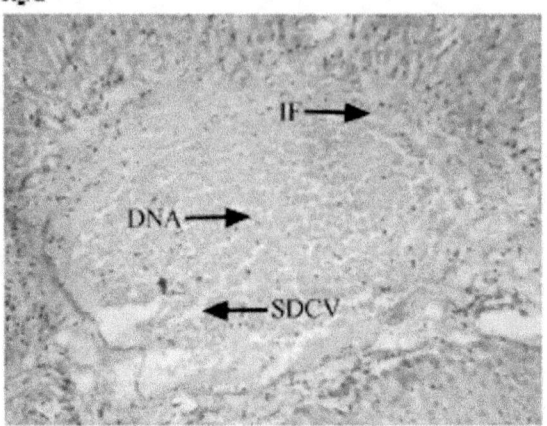

Fig. F

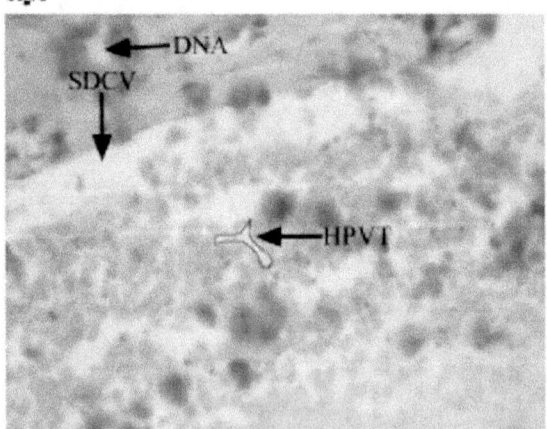

LEGEND FOR FIGURES

Plate 5.1

Fig. G: 72 hrs of Snake *Naja naja* venom administrated rat liver showing Sinusoidal Haemorrhage (SH), Focal Necrotic Areas (FNA), and Amyloid Precipitation (AP), Hepatocyte Cell With Prominent Nucleus (HCPN).   H & E. 400 X.

PLATE – 5.1

Fig. G

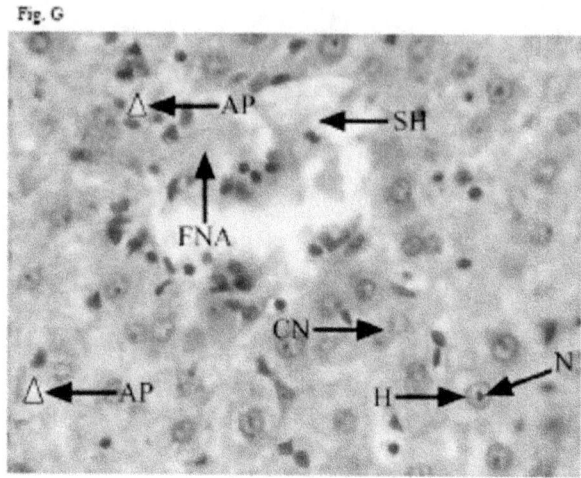

LEGEND FOR FIGURES

Plate 5.2

Fig. A: Control rat kidney showing Glomerulus (G), Glomeruli with Stalk (GS), Bowman's Capsule (BC), proximal Convoluted Tubule (PCT), and Distal Convoluted Tubule (DCT). H & E. 100 X.

Fig. B: 24 hrs of Snake *Naja naja* venom administrated rat kidney showing Slight Thickening Of Bowman's capsule (STBC), and Infiltration (I), Glomeruli Stock (GS), Bowman's Capsule (BC) and Glomeruli (G),Congestion of Blood Vessels (CBV). H & E. 100X.

PLATE – 5.2

Fig. A

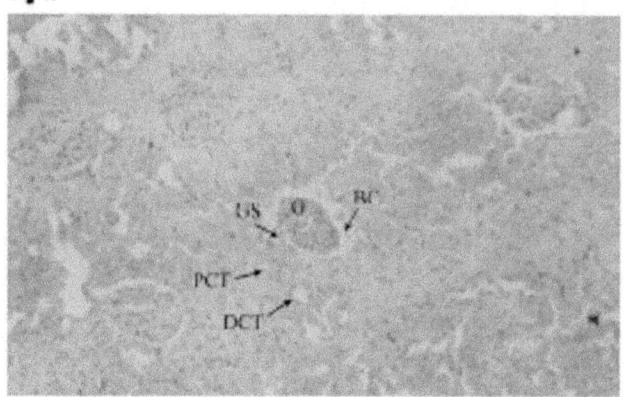

Fig. B

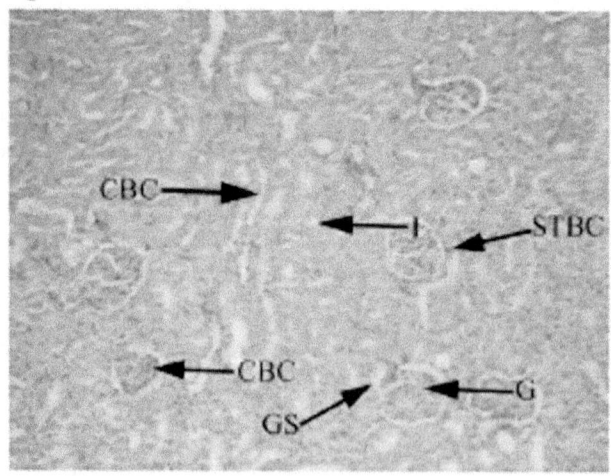

LEGEND FOR FIGURES

Plate 5.2

Fig. C: 24 hrs Snake *Naja naja* venom administered rat kidney showing hypertrophy of Bowman's Capsule (HBC) and Fragmentation of Glomerulus (FG), Congestion (C) and Slight Thickening of Bowman's Capsule (STBC). H & E. 100X.

Fig. D: 24 hrs snake *Naja naja* venom administered rat kidney showing atrophy of Glomerulus (AG), and Hypertrophy of Bowman's Capsule (HBC). H & E. 100X.

PLATE – 5.2

Fig. C

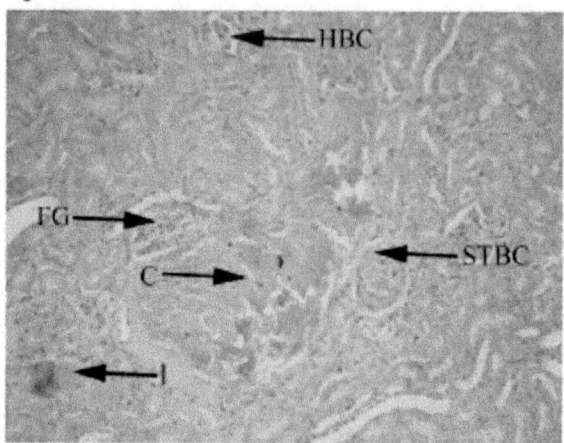

Fig. D

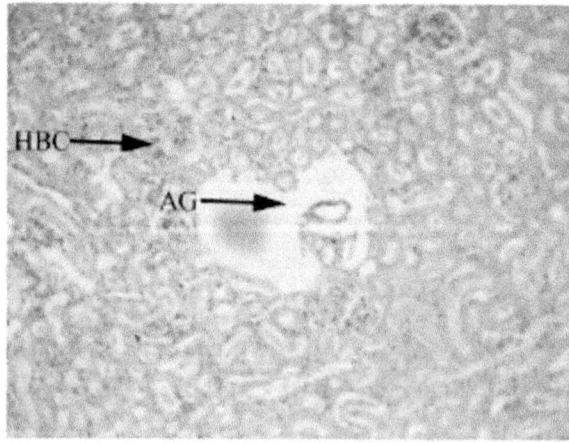

LEGEND FOR FIGURES

Plate 5.2

Fig. E: 72 hrs of snake *Naja naja* venom administered rat kidney showing Fragmentation of Glomerulus (FG), and Increased Size of the Lumen (ISL), Hypertrophy of Bowman's Capsule (HBC), Congestion in Blood Vessel(CBC). H & E. 400 X.

Fig. F: 72 hrs of snake *Naja naja* venom administered rat kidney showing Blood Vessel Congestion (BVC). H & E. 400 X.

PLATE – 5.2

Fig. E

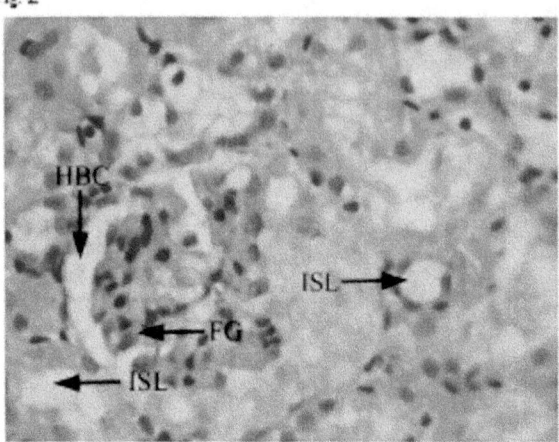

Fig. F

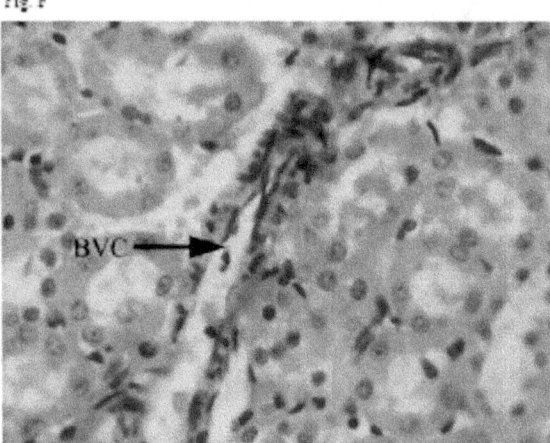

LEGEND FOR FIGURES

Plate 5.3

Fig. A :   Normal structure of Control rat Cerebral Cortex showing Molecular Layer (ML), Outer Granular Layer (OGL), Outer Pyramidal Layer (OPL), with Betz cells (BZ), Stellate Neurons (SN), Glial Cells (GC), Pyramidal Cells (PC) and the neurons with perikaryon. H & E. 400 X.

Fig. B :   Cortex of rat under 4X of snake *Naja naja* venom administered rat Brain showing Molecular Layer (ML), Outer Granular Layer (OGL), Outer Pyramidal Layer (OPL), Stellate Neurons (SN), Glial Cells (GC), Pyramidal Cells (PC). H & E. 40 X.

PLATE – 5.3

Fig. A

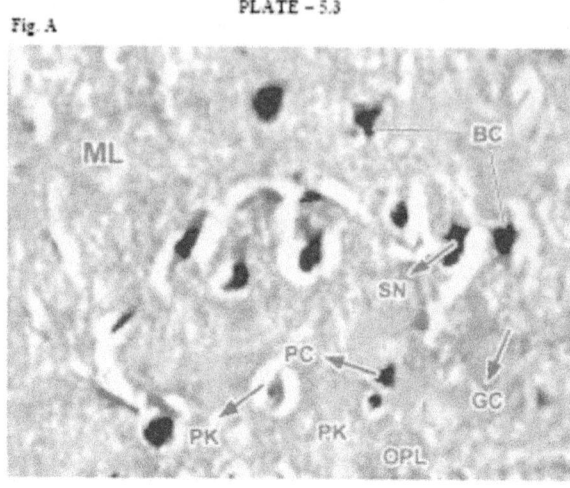

Fig. B

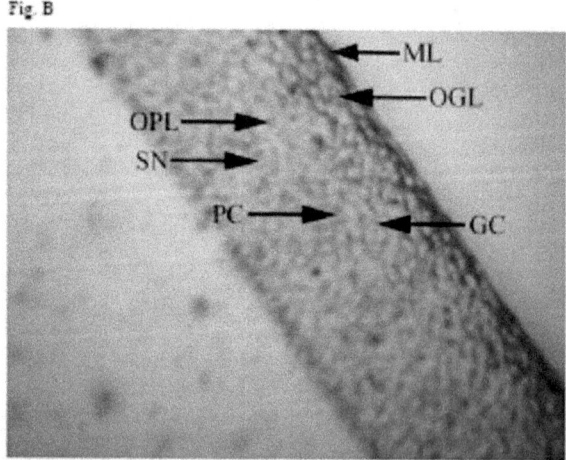

LEGEND FOR FIGURES

Plate 5.3

Fig. C :   24 hrs of snake *Naja naja* venom of rat brain cortex showing
Molecular Layer (ML), Granular Layer (GL), Outer Pyramidal
Layer (OPL), Stellate Neurons (SN), Mild Blood Vessel
Congestion (MBVC), Mild Vacuolation (MV), Betz Cells
(BZC), Pyramidal Cells (PC). H & E. 100 X.

Fig. D :   24 hrs of snake *Naja naja* venom of rat brain cortex showing
Picknosis (PK), Pyramidal Cells (PC), Glial Cells (GC), Betz
Cells (BZC). H & E. 400 X.

PLATE – 5.3

Fig. C

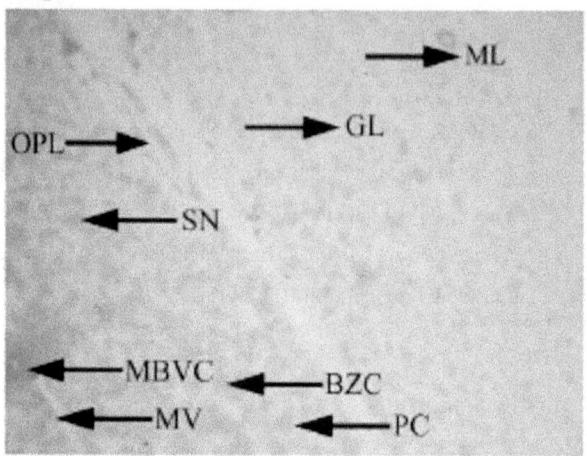

Fig. D

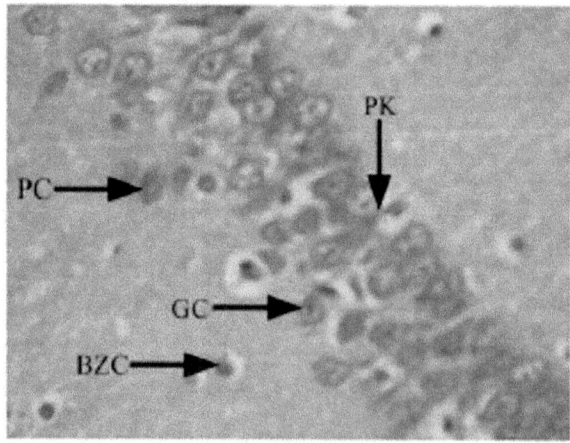

LEGEND FOR FIGURES

Plate 5.3

Fig.E :   Control cerebellum showing a prominent   Molecular Layer
(ML),Granular Layer (GL), and Congestion of Purkinje Cells
(CPJC). H & E. 100 X.

Fig.F: 24 hrs of Snake *Naja naja* venom administered rat brain
cerebellum showing Molecular Layer (ML), Congestion of
Purkinje Cells (CPJC), Necrosis in the Layer of Purkinje Cells
(NPJC), Severe   Degenerative Changes in the Granular Layer
(SDCGL). H&E.100 X.

PLATE – 5.3

Fig. E

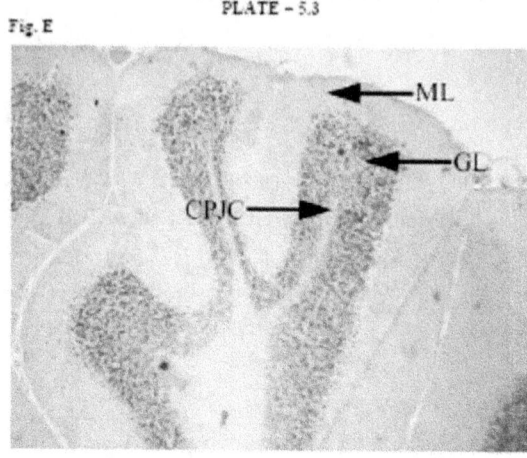

Fig. F

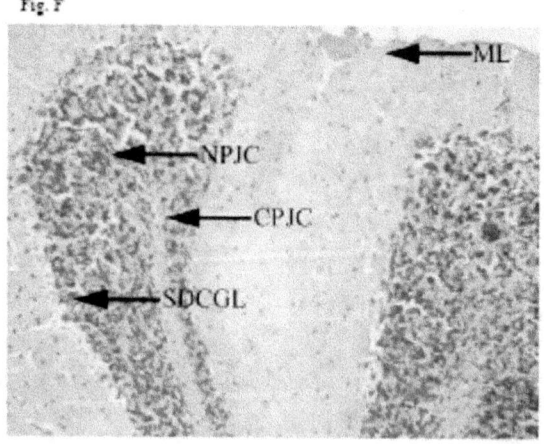

LEGEND FOR FIGURES

Plate 5.3

Fig.G :   72 hrs of Snake *Naja naja* venom administered rat brain showing Blood Vessel Congestion (BVC), Severe Degenerative Changes in Granular (SDC in GC), Severe Degenerative Changes in Purkinje Cell (SDC in PJC), Congested Glial Cells (CGC),Vacuole Formation (VF). H & E. 400X.

Fig.H :   Control hippocampus of rat brain showing Cornu Ammonis Layer (CA) with different neurons viz., Spindle Neurons (SN), Stellate Neurocytes (SNC), Pyramidal Neurons (PN), Glial Cells (GC) and Neurofibrillar Network(NFN). H & E. 400 X.

PLATE – 5.3

Fig. G

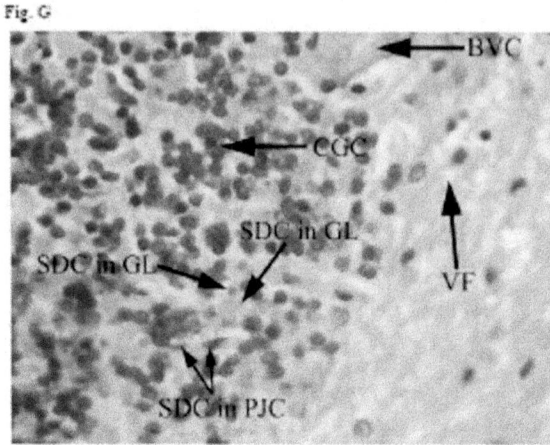

Fig. H

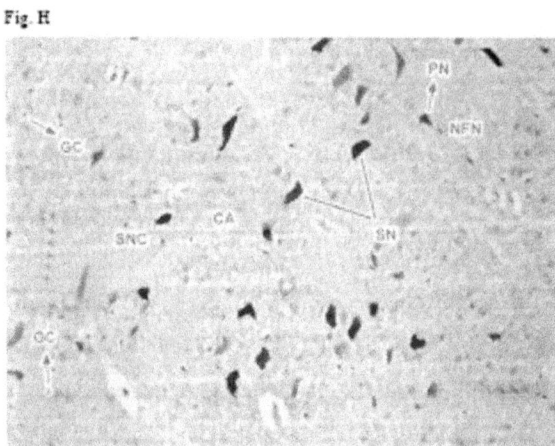

LEGEND FOR FIGURES

Plate 5.3

Fig. I :   24 hrs of Snake *Naja naja* venom administered rat brain hippocampus showing Cornu Ammonees Layer (CA1), Cornu Ammonis Layer (CA3), Granular Cells (GC). H & E. 100X.

Fig. J :   72 hrs of Snake *Naja naja* venom administered rat brain hippocampus showing Severe Necrosis in Glial Cells (SN in GC), Severe Necrosis of Neurofibrillar Network (SNNFN),Vacuole Formation (VF), Blood Vessel Congestion (BVC). H & E. 400X.

PLATE – 5.3

Fig. I

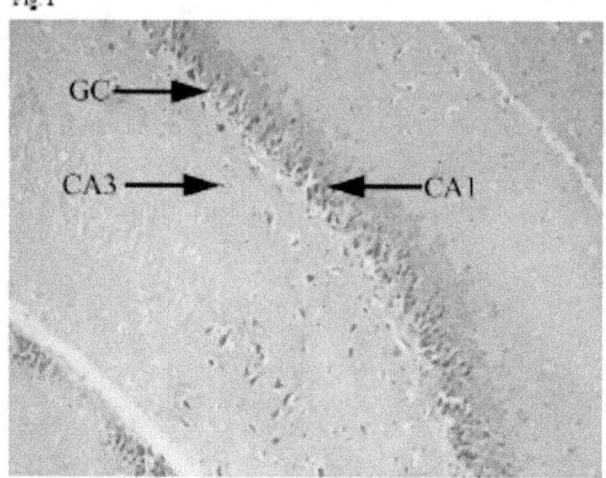

Fig. J

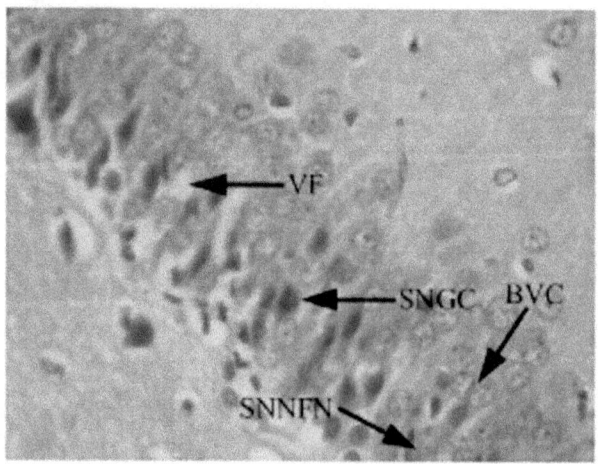

LEGEND FOR FIGURES

Plate 5.4

Fig. A: Control rat Heart Showing Cardiac Muscle Fibers with Rounded Nucleus (RN). H & E.100X.

Fig. B: 24 hrs Snake *Naja naja* venom administered rat heart showing Slight Infiltration (SI) and Blood Vessel Congestion (C). H&E. 100 X.

PLATE – 5.4

Fig. A

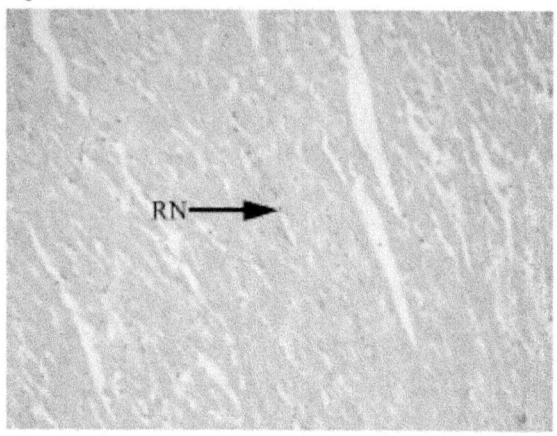

Fig. B

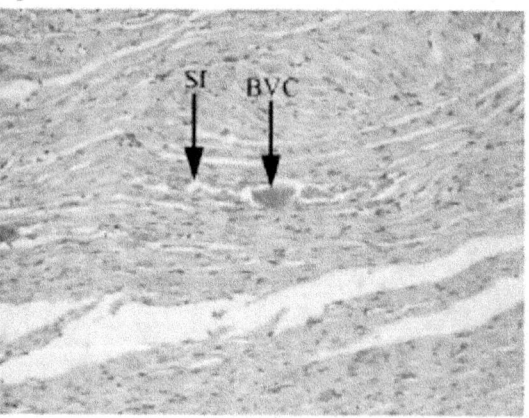

LEGEND FOR FIGURES

Plate 5.4

Fig. C :   24 hrs   of Snake   *Naja naja* venom administered rat heart showing Rounded Nucleus (RN). H & E 400 X.

Fig. D :   72 hrs   of Snake *Naja naja* venom administered rat heart showing   Diffused Areas of Haemorrhage (DAH), Rounded Nucleus (RN).   H & E. 400X.

PLATE – 5.4

Fig. C

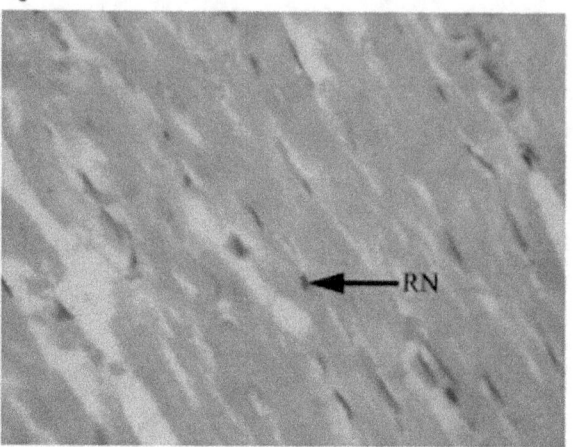

Fig. D

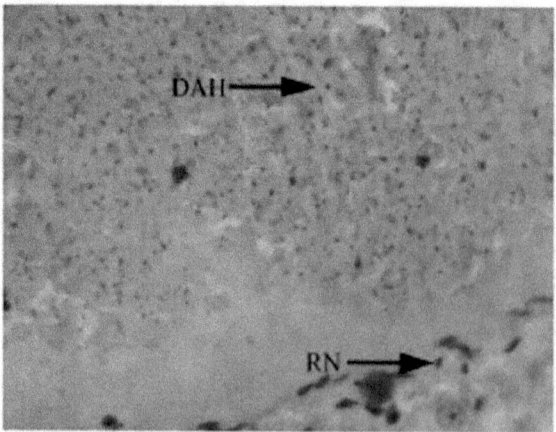

LEGEND FOR FIGURES

Plate 5.4

Fig. E :   72 hrs of Snake *Naja naja* venom administered rat heart showing   Blood Vessel Congestion (BVC), and Slight Infiltration (SI), Rounded Nucleus (RN).   H & E. 400 X.

Fig. F :   72 hrs of *Naja naja* venom administered rat heart showing Degenerated Muscle Fibre (DMF). H & E. 400 X.

PLATE – 5.4

Fig. E

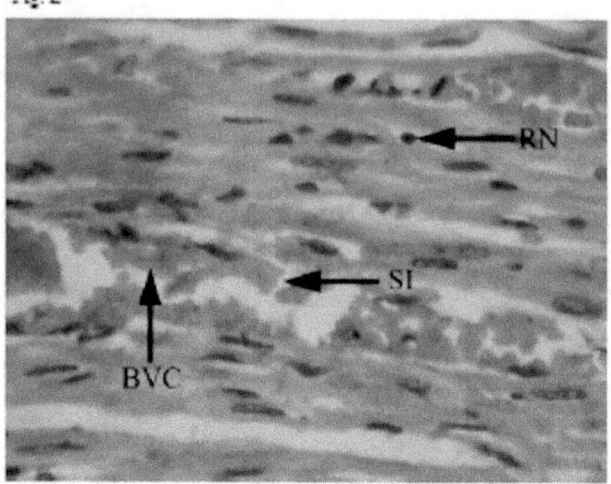

Fig. F

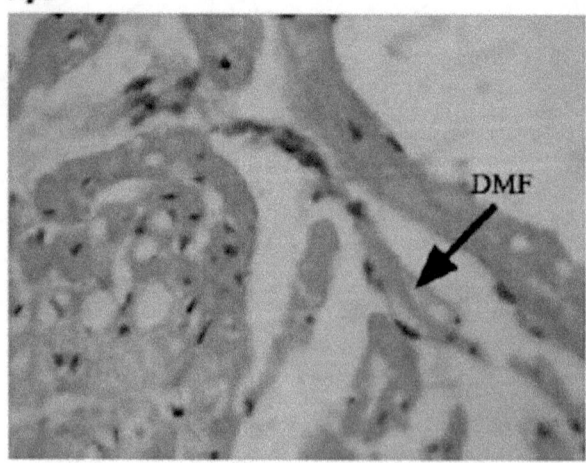

Summary and Conclusions

The snake venom is considered as a major and serious problem of today's world. Continuous exposure to poisonous snake venom results in their accumulation in various components of the tissues. Presence of these snake venom above the safe level in the tissues will pose serious threat to human beings. The serious effects of snake venom on human life and health is indirectly probed through this investigation.

❖ The present study was aimed to elucidate the changes in haematological profiles, enzymes of protein metabolism, detoxification mechanisms and cytoarchitectural studies in Albino rat of snake *Naja naja* venom envenommation.

❖ Wistar strain Albino rats were selected as the experimental animal and the snake *Naja naja* venom is supplied by Irula Snake Catchers, Tamilnadu. Sodium chloride was used as the vehicle for administering snake *Naja naja* venom. The route of administration is by oral intubation method.

❖ The toxicity evaluation of snake *Naja naja* venom was done by probit method of Finney (1971) and the LD_{50} value thus determined was 38.4 mg/kg body weight / 48 hrs. Since the venom is highly poisonous, sub lethal dose was selected as $1/50^{th}$ of LD_{50} i.e., 0.768 mg/ Kg body weight. The graphical representation of percent mortality versus log concentration and probit mortality versus log concentration showed a typical sigmoid curve and straight line respectively, and were in agreement with the principle of probit analysis (Finney, 1971).

❖ Albino rats were divided into 4 groups, each group having 10 animals. Group I was treated as control, where as II, III and IV groups were treated as experimental groups with increase on time schedules i.e., 24 hours, 48 hours, 72 hours respectively with an interval of 24 hours, dose was given. The animals were sacrificed by cervical dislocation and the blood was collected for hematological study, and different tissues viz., liver, kidney, brain, heart were isolated for histological studies and the remaining tissues were stored at -80 °C for further biochemical investigations.

❖ Haematological profiles were altered in Snake *Naja naja* venom intoxicated rats. A gradual decrease in RBC, Hb, PCV, MCV, MCH, MCHC, platelet count, serum creatinine and increase in WBC, were observed. In differential count also, a gradual decrease in neutrophils, serum total proteins and increase in lymphocytes and monocytes, eosinophils, basophils, cholesterol, triglycerides, alkaline phosphatase, gamma glutamyl transferase, and electrolytes such as sodium, potassium, calcium, phosphorous, amylase, lactate dehydrogenase were observed. The change in haemogram indicates that the snake *Naja naja* venom alters the biochemical pathways and causes cellular damage.

❖ Total proteins were found decreased, while the free amino acid levels were elevated in different tissues under snake *Naja naja* venom envenomation. The decreased level of total proteins, elevated level of free amino acids were progressive on time schedules in the present study. The decrease in protein content indicates proteolysis leading to elevation in total free amino acids content.

❖ Protease activity in the different tissues were found increased upon exposure to snake *Naja naja* venom. The increased protease activity level with increase of time schedule suggests that there is a greater protein degradation to meet the cellular demands.

❖ Aminotransferase activities (AST & ALAT) were increased in snake *Naja naja* venom administration. Maximum elevation of AST and ALAT activities were observed in 72 hrs of time interval. The elevation in AST

184

and ALAT activities suggests the enhanced feeding of amino acids in to the energy cycle.

❖ The glutamate dehydrogenase activity (GDH) was found to be elevated in all the tissues of snake *Naja naja* venom intoxicated rats. The elevated GDH activity levels indicate its contribution to ammonia production and glutamate oxidation during snake *Naja naja* venom toxicity. The elevated free amino acid levels and their subsequent transamination towards the formation of glutamate leads to the consequent oxidative deamination reaction through GDH and also helps in supplying keto acids to TCA cycle for energy production.

❖ Ammonia levels were elevated in all the tissues of snake *Naja naja* venom intoxicated rats. In tissues, the increased protein catabolism leads to the elevation of amino acids level, which caused to the increase of ammonia level by transamination and deamination processes. Thus the large amount of ammonia was accumulated in all the tissues.

❖ Urea levels were increased in liver tissue might be due to activation of urea cycle. The presence of urea in extra-hepatic tissues might be due to the vascular mobilization and translocation from liver observed in the tissues of rats exposed to sublethal doses of snake *Naja naja* venom. The elevation in urea levels was in consonance with increased proteolytic activity, enhanced transamination and elevated ammonia levels during snake *Naja naja* venom toxicosis. Increased levels of urea under snake *Naja naja* venom stress, reveal that, the rats might have been adapted to the biosynthesis of urea as a major pathway of detoxification of ammonia.

❖ Snake *Naja naja* venom affect the major vital organs like liver, kidney, brain, heart and the DNA banding pattern was found to alter in envenomated rats. The DNA banding pattern revealed that the snake *Naja*

185

naja venom exposed to liver, kidney, brain, heart interact with the DNA intact bands there by causing DNA fragmentation on time dependent manner seen during the experimental period.

❖ Xanthine oxidase (XOD) activity levels were elevated in all the tissues with increase on time schedules. The elevated levels of XOD activity indicates the over production of oxygen free radicals in response to the snake *Naja naja* venom toxicity.

❖ Superoxide dismutase (SOD) activity was assayed to observe the levels of detoxification of superoxide anion radicals and SOD activity was decreased on time schedules. The SOD activity reflects the oxidative status of tissues, there by the cells of different tissues were damaged by free radicals. This oxidative stress produces depletion of SOD activity.

❖ Catalase activity (CAT) was estimated to assess the hydrogen peroxide reduction potential of different tissues. The catalase activity was also depleted like SOD with an increase on time intervals of snake *Naja naja* envenomation. It is due to cell damage resulting in oxidative stress leading to the depletion of catalase activity.

❖ In the present investigation, histopathological studies were carried out by light microscopy, to elucidate the toxic potential of the snake venom in rats exposed to sub lethal dose of snake *Naja naja* venom. Severe histopathological changes were observed in 72 hrs of envenomated rats.

❖ The experimental animals showed marked pathological changes in different tissues which include infiltration, congestion, necrotic changes, cellular swelling, degenerative changes, hypertrophy of bowman's capsule, atrophy of glomerulus, fragmentation of glomerulus, hyaline deposits in the lumen, increased size of lumen in proximal and distal convoluted tubules, hyperemia, and increased connective tissue by light microscopy.

186

From the above study involving haematological, protein metabolism, electrophoresis, and detoxification mechanisms and histopathological observations, it is concluded that these parameters play crucial role in different metabolism and animal survivability. Estimation of biochemical components and enzymes in envenomated rats gives a clear picture of organs effected by snake *Naja naja* venom in general and confirmed their necrotic potency against vital organs especially in liver, kidney, brain and heart. This study would be of importance in characterization of possible proteins in snake *Naja naja* venom as well as understanding their action in the process of development of a specific antivenom. The pathophysiological condition in envenomted rats would extrapolate to know the pathological condition and severity of snake bites in human beings.

Bibliography

Abdel-Nabi, I.M. (1993). Effect of crude *Cerastes cerastes* venom and fraction B on the clinical biochemical parameters of white rat. *Journal of the Egyptian German Society of Zoology.* **10 (A)** : 315-326.

Abdel-Nabi, I.M., Raafat, A. and El-Shamy, H.I. (1997). Biological effects of intraperitoneal injection of rates with the venom of the snake *Echis carinatus. Egyptian Journal of Zoology.* **29** : 195-205.

Adali, M.M., Inal-Erden, A., Akalin. and Efe, B. (1999). Effects of propylthiouracil, propranol and vitamin E on lipid peroxidation and antioxidant status in hyperthyroid patients. *Clin. Biochem.* **32**: 363-367.

Aebi, H. (1984). Catalase. *Methods Enzymol.* **105** : 125-126.

Ahn, M.Y., Lee, P.M., Kim, Y.S. (1997). Characterization and cytotoxicity of L-amino acid oxidase from the venom of king cobra (*Ophiophagus hamah*). *Int. J. Biochem, Cell Biol.* **29 (6)** : 909 - 911.

Al-Sadoon, M.K. and Haffor, A.S.A. (2005). Effects of *Cerastes cerastes gasperettii* venom on hepatocyte mitochondria ultrastructure and blood cells count. *J.Med.Sci.* **5(4)**: 253-259.

Alam, J.M. and Ali, S.A. (1998). Comparative biochemical and biological studies on venoms of cobra *Naja naja* snakes from different localities. *Pakistan J Zool.* **50** : 205-206.

AL-Asmari, A.K. (2005). Pharmacological characterization of rat paw edema induced by *Naja naja arabica* venom. *J. venom Animal Toxins.* **11** : 51-67.

AL-Jammaz, I., Fahim, A. and Attia, M.A. (1993). Comparative physiological and biochemical investigation on the venom effect of two Elapidae on rats. I-carbohydrate metabolism *Proc. Zool. Soc. A.R.E.* **24** : 145-151.

AL-Jammaz, I., Al-Sadoon, M.K., Attia, M.A. and Fahim, A. (1994). Effect of *Walterinnessia aegyptia* venom on the serum and tissue metabolites and some enzyme activities in male albino rats-III. Effects on lipid metabolism and two dehydrogenases. *Journal of King Saud University (Science).* **6 (2)** : 207-215.

Alka Gupta., Amita Gupta and Girija Shukla, S. (1999). Effects of neonatal quinolphos exposure and subsequent withdrawl on free radical generation and anti oxidative defenses in developing rat brain. *Journal of applied toxicology.* **18 (1)** : 71-77.

Aung-Khin, M. (1978). Histological and ultrastructural changes of the kidney in renal failure after viper envenomation. *Toxicon.* **16**: 71-75.

Awasthi, M., Shaw, P., Dubale M.S. and Gadhia, P. (1984). Metabolic changes induced by organophosphates in the piscine organs. *Environ. Res.* **35**: 320-325.

Babij, P., Mathews S.M. and Rennie, M.J. (1983). Change in blood ammonia, lactate and amino acids in relation to workload during bicycle exercise in man. *Eur. J. Physiol.* **50** : 405 - 411.

Babsky, E., Khodorov, B., Kositsky, G. and Zubkov, A. (1985). In : **Human physiology**. (Edited by Babsky, E.B.). MIR. Publishers. Moscow, pp.56.

Baccari, E. (1949). A method for reducing the number of pharmacological assays. *Journal of Nature.* **165** : 168.

Bahman Maroufi., Kaboudanian Sussan Ardestani., Amina Kariminia., Hussain Naderimanesh. (2005). The effect of vitamin E on splenocytes apoptosis of gamma –irradiated BALB/C mice . *Iranian Journal of Allergy , Asthma and Immunology.* **4 (2)** : 77-82.

Banister, E., Rajendra, W. and Mutch, B.J.C. (1988). Ammonia as an indicator of exercise stress : Implications of recent findings to sports medicine. *Sports. Medicine.* **2** : 34-36.

Barker, R.A., Ratcliffe, E., McLaughlin, M., Richards, A. and Dunnett, S.B. (2000). A role for complement in the rejection of porcine ventral mesencephalic xenografts in a rat model of parkinson's disease. *J.Neurosci.***20**: 3415-3424.

Barrett, A.J. (1977). Introduction to the histology and classification of tissue proteinases. In : **Proteinases in mammalian cells and tissues.** (Edited by Barrett, A.). Amsterdam, pp. 1-55.

Batra, S., Singh, S.P., Srivastava, V. M. L. and Chatterjee, R. K. (1989). Xanthine oxidase, superoxide dismutase and catalase and lipid peroxidation in mastomys netalensis. Effect of dipetaloema viteae infection. *Ind. J. Exp. Biol.* **27**: 1067-1070.

Beachamp, C. and Fridovich, I. (1971). Superoxide dismutase improved assay and an assay applicable to PAGE. *Analyt. Biochem.* **44** : 276-287.

Bergmeyer, H.V. (1965). In: **Methods of enzymatic analysis.** (Edited by Bergmeyer, H.V.). Academic Press, New York, pp. 401.

Bergmeyer, H.V. and Bernt. (1965). In : **Methods of enzymatic analysis.** (Edited by Bergmeyer, H.V.) Academic Press, New York, pp. 401.

Bessman, S.P. and Pal, N. (1976). The Kreb's cycle depletion theory of hepatic coma. In : **The urea cycle.** (S. Grisolia, R. Bagnena, F. Mayer Eds.). John Wiley and Sons, Inc. pp. 494.

Bhargava, D. (1982). Fatigue induced changes in carbohydrate and protein metabolism and their modulation by pyridoxal-5-phosphate in the gastrocnemius muscle of *Bufo melanostictus.* Ph.D. Thesis, Sri Venkateswara University, Tirupati, India.

Bieber, A.L. (1979). Metal and nonprotein constituents in snake venoms. In : **Snake venoms, Handbook of Experimental Pharmacology. 52** : (Edited by Lee, C.Y.). Springer – Verleg, Berlin, pp. 295.

191

Bradbury, S.P., Mc Kim, J.M. and Coasts, J.R. (1987). Physiological responses of rainbow trout, *Salmo gairdneri* to acute fenvalerate intoxication. *Pestic. Biochem. Physio.* **27** : 275 - 288.

Braganca, B.M. and Quastel, J.H. (1953). Enzyme inhibitions by snake venoms, *Biochem.J.* **53**: 88-94.

Braganca, B.M., Sambray, Y.M. and Sambray, R.Y. (1970). Isolation of polypeptide inhibitor of phospholipase A from cobra venom. *Europ. J. Biochem.* **13** : 410-416.

Braud, S., Bon, C. and Wisner, A. (2000). Snake venom proteins acting on hemostasis. *Biochemie.* *82:* 851-859.

Braunstein, A.E. (1973). Amino group transfer In : **The enzymes.** (Edited by Boyer, P.D.), 3[rd]edition. Academic Press , New York, pp. 379-481.

Broad, A. J., Sutherland, S. K., Coulter, A.R. (1979). The lethality in mice of dangerous Australian and other snake venoms. *Toxicon* **17** : 664-667.

Brown, V.K.H. (1985). Acute toxicity testing, In: **Animals and Alternatives in Toxicity Testing,** (Balls M, Riddell RJ and Worden AM Eds.). *Academic Press*, London, pp.474.

Bruce, R.D. (1985). An up and down procedure for acute toxicity testing. *Fundam. Appl. Toxicology.* **5** : 151-157.

Cabre, M., Comps, J., Paternain, J.L., Ferre, N. and Joven, J. (2000). Time course of changes in hepatic lipid peroxidation and glutathione metabolism in rats with carbon tetrachloride induced cirrhosis. *Clin. Exp. Pharmacol. Physiol.* **27-9** : 694-699.

Calabrese, A., Thurberg F.P. and Gould, E. (1977). Effects of cadmium, mercury and silver on marine animals. *Mar. Fish. Rev.* **39** : 5-11.

Calmette, A., Saenz, A. and Costil, L. (1933). Effects du venimde cobra surles greffes conceruses et sur le cancer spontane (adeno carcinoma) de la souris, *CRA cad sci.* 197 : 205.

Calvete, J.J., Escolano, J. and Sanz, L. (2007a). Snake venomics of *Bitis* species reveals large intragenus venom toxin composition variation: application to taxonomy of congeneric taxa. *J Proteome Res.* 6 : 2732-2745.

Calvete, J.J., Juarez, P. and Sanz, L. (2007b). Snake venomics. strategy and applications. *J. Mass Spectrom.* **42** : 1405-1414.

Calvete, J.J., Marcinkiewicz, C. and Sanz, L. (2007c). Snake venomics of *Bitis gabonica gabonica*. Protein family composition, subunit organization of venom toxins, and characterization of dimeric disintegrins bitisgabonin-1 and bitisgabonin-2. *J. Proteome Res.* **6:** 326-336.

Campbell, J.W. (1973). Nitrogen excretion In : **Comparative Animal physiology** (Edited by Prosser, CI.). Saunders Co., London, pp. 279 - 316.

Cartson, S. (1962). Test methods and toxicity considerations under the federal hazardous substances labeling. *Act Proc. of 49th Ann. Meeting Chem. Spec. Assam.* New York. pp. 12.

Chang, C.C. (1979). The action of the snake venoms on nerve and muscle In: **Handbook of Experimental pharmacology**, (Edited by Lee, C.Y.). Springer –Verlag. pp. 310-312.

Chang, L.S., Chung, C., Liou, J.C., Chang, C.W. and Yang, C.C. (2003). Novel neurotoxins from taiwan banded krait *(bungarus multicinctus)* venom: purification, characterization and gene organization. *Toxicon.* **42** : 323–330.

Chugh, K., Aikat, B.K., Sharma, B.K., Dash, S.C., Mathew, M.T. and Das, K.C. (1975). Acute renal failure following snakebite. *The American Journal of Tropical medicine and Hyg.* **24 (4)** : 692-697

Ciutat, D., Caldero, J., Oppenheim, R.W., Esquerda, J.E. (1996). Schwann cell apoptosis during normal development and after axonal degeneration induced by neurotoxins in the chick embryo. *Journal of neuroscience.* **16** : 3979–3990.

Colowick, S.P. and Kaplan, (1957). In: **Methods of Enzymology**, *Academic Press*, New York, pp. 501.

Cousin, X. and Bon, C. (1999). Acetyl cholinesterase from snake venom as a model of its nerve and muscle counterpart. *J.Nat.Toxins.* **8 (2)**:285-294.

Covacevich, J. and Wombey, J. (1976). Recognition of *Parademansia kmicrolepidotus McCoy* (Elapidae), a dangerous Australian snake. *Proc, R. Soc. Qd.* 87-29.

De Silva, H.J., Ratnatunga, N., De Silva, U., Kularatne, W.N., Wijewickrema, R. (1992). Severe fatty change with hepatocellular necrosis following bite by a Russell's viper. *Trans. R. Soc. Trop Med.Hyg.* **86 (5)** : 565.

Dellacorte, E. and Stripe, F. (1972). The regulations of liver xanthine oxidase. Involvement of thiol groups in the conversion of the enzyme activity from dehydrogenase (type-D) into oxidase (type-O) and purification of the enzyme. *Biochem. J.* **126** : 739-745.

Dieter, H., Kobensterin, R. and Sund, H. (1981). Studies of glutamate dehydrogenase. The interaction of ADP, GTP and NADPH in complexes with GDH. *Environmental Journal of Biochemistry* **15**: 217-226.

Dolly, J.O., Black, J.D., Black, A.R., Pelchen-Matthews, A. and Halliwell, J.V. (1986). Novel roles of neural receptors for inhibitor and facilitatory toxins. In : **Natural Toxins, Animal, Plant, Microbial**, (Edited by Harris, J.B.). Clarendon Press, Oxford, pp. 237.

Eaton, D.L. and Klaassen, D. (1996). Principles of toxicology. In : **The Basic Science of Poisons** (Edited by Casarett and Doulls). McGraw-Hill, New York, pp.13-33.

Edwards, C.A. (1973). Effect of pesticide residue on soil invertebrates and plants. *Br. Ecol. Soc. Symp.* **5** : 239.

Eldadah, B.A., Yakovlev, A.G. and Faden, A1. (1996). A new approach for the electrophoretic detecition of apoptosis. *Nucleic Acids Research.* **24 (20)** : 4092-4093.

Elliott, W.B. (1978). Chemistry and immunology of reptilian venoms. In: **Biology of the Reptilia** (Edited by Gans, C.). *Academic Press,* New York, pp.163.

Fahim, A. (1998). Biological effects of the viper *Bitis arietans,* crude venom on albino rats. *Egypt.J. Zool.* **30** : 35-54.

Fattman, C.L., Schaefern, L.M. and Oury, T.D. (2003). Extracellular superoxide dismutase in biology and medicine. *Free Radic. Biol. Med.* **35-3** : 236-256.

Ferrari, A., Venturino, A. and Pechende Angelo, M. (2007). Effect of carbamyl and azinfos methyl juvenile rainbowtrout (*Oncorynchu smykis*) detoxifying enzymes. *Pesticide biochemistry and physiology.* **88 (2)**: 134-142.

Finney, D.J. (1971). In : **Probit analysis**, III Edition, Cambridge University press, London, **20**.

Folin, O. and Wu, H. (1991).Estimation of serum creatinine. *J. Biol, Chem.* **38,** 81.

Fox, J.W. and Serrano, S.M. (2008): Exploring snake venom proteomes: multifaceted analyses for complex toxin mixtures. *Proteomics.* **8 (4)** : 909-920.

Fry, B.G. (2005). From genome to venome, molecular origin and evolution of the snake venom proteome inferred from phylogenetic analysis of toxin sequences and related body proteins. *Genome Res.* **15** : 403-420.

Fry, B.G. Vidal, N., Norman, J.A., Vonk, F.J., Scheib, H., Ramjan, S.F., Kuruppu, S., Fung, K., Hedges, S.B.and Richardson, M.K. (2006). Early evolution of the venom system in lizards and snakes. *Nature.* **439**: 584 - 588.

Fukuda, K., Mizuno, H., Atoda, H. and Morita, T. (2000). Crystal structure of flavocetin-A, a platelet glycoprotein Ib-binding protein, reveals a novel cyclic tetramer of C-type lectin-like heterodimers. *Biochemistry.* **39** : 1915-1923.

Geiger, R. and Kortmann, H. (1977). Esterolytic and proteolytic activities of snake venoms: and their inhibition by proteinase. *Toxicon.* **15**: 257.

Giardiano, F. J. (2005). Oxygen, oxidative stress, hypoxia and heart failure. *J. Clin. Invest.* **115** : 500-508.

Gitter, S., Moroz-Perlmutter, C., Boos, J.H., Liveni, E., Rechnic, J., Goldbtum, N . and De Vries, (1962). A Studies on the snake venoms of the near east: *Walterinnesia uegyptia and Pseudocerastes fieldii, J. Trop. Med. Hyg.* **11**: 861-875.

Goldberg, A.L. and Dice, J.F. (1974). Intra cellular protein degradation in mammalian and bacterial cells. *Annu. Rev. Biochemistry.* **43**: 835-869.

Gomorri, G. and Bab, J. (1942). Estimation of serum phosphorous. *Clin.Med.* 27-35**(5)**.

Gould, E., Collier, R.S., Karolous J.J. and Givenus, S. (1976). Heart transaminases in the rock crab, *Cancer irroratus* exposed to cadmium salts. *Bulletin of Environmental Contamination and Toxicology.* **15** : 635-643.

Grainde, B. and Seglen, P.O. (1981). Effects of amino acid analogues on protein degradation in rat hepatocytes to resin acids in Mediterranean mussels. *Eco Toxicol-environ. Saf.* **61 (2)** : 221-229.

Groten, J.P., Schoen, E.D., Kuper, E.F., Von Bladeren, P.J., Van Zorge J.A. and Feron, V.J. (1997). Subacute toxicity of a mixture of nine chemicals in rats, deducting interactive effects with a 2 level factorial design. *Fundam. Appl. Toxicology.* **36(1)** : 15-29.

Gupta, R.C., Prabhakar Rao, Malik, J.K., Singh, R.V., Verman, P.N. and Paul, B.S. (1980). Influence of fenitrothion on *in vitro* incorporation of acetate 1-140 in liver lipids and various tissue enzymes in rats. *J. Nuclear Agric. Biol.* **9** : 25-28.

Gutierrez, J.M., Lomonte, B. and Cordas, L. (1986). Isolation and partial characterization of a myotoxin from the venom of the snake *Bothrops mommifer. Toxicon.* **24** : 835.

Halliwell B. and Gutteridge J.M.C.(1990). The antioxidants of human extracellular fluids. *Arch Biochem Bio phys.* **280** : 1-8.

Harris, H.F. (1900). On the rapid conversion of haematoxylin into haemation in staining reaction. *J. Appl. Microse. Lab. Meth.* **3** : 777.

Hawkins, W.W., Speck, E. and Leonard, V.G. (1954). Variation of the hemoglobin level with age and sex. *Blood.* **9(10)** : 999-1007.

Henson, P.H. and Cochrane, C.G. (1975). The effect of complement depletion on experimental tissue injury. *Ann Ny Acad Sci.* **256** : 426.

Humason, G.L. (1972).In : **Animal tissue technique**, III Edn., W.H. Freeman and Company, San Francisco.

Ibrahim, A. and AL-Jammaz. (2002). Effects of envenomation by Cerastes vipera crude venom on plasma and tissue metabolites of rats. Kuwait *J.Sci.Eng.* **29(1):** 111-119.

Ibrahim, Al-Jammaz, I., Al-Sadoon, M.K., Attia, M.A. and Fahim, A. (1994). Effect of *Walterinnessia aegyptia* venom on the serum and tissue metabolites and some enzyme activities in male albino rats-III. Effects on lipid metabolism and two dehydrogenases. *Journal of King Saud University (Science).* 6(2): 207-215.

Ibrahim. A. and AI-Jammaz. (2003). Physiological Effects of LD_{50} of *Echis coloratus* crude venom on rat at different time intervals. *J.King Saud Univ.* 15 (2): 135-143.

Ismail, M., Gumma, K.A., Osman, O.H. and El-Asmar, M.F. (1978). Effect of *Buthus minax* (L-Koch) scorpion venom on plasma and urinary electrolyte levels. *Toxicon.* 16: 385-392.

Iwanaga. S. and Suzzuki, T. (1979). Enzymes in snake venoms. In: **Handbook of Experimental Pharmacology 52:** (Edited by Lee.C.Y.). Springer – Verleg, Berlin, pp. 61.

Jacoby, W.B. (1980). Detoxification enzymes, In: **Enzymatic basis of detoxification**, Vol.II, (Edited by Jacoby W.B.). Academic press, New York, pp.124.

Jagadeesan, G. and Mathivanan, A. (1999). Organic constituent changes induced by 3 different sublethal concentrations of mercury and recovery in the liver tissue of *Labeo rohita* finger lings. *Poll. Res.* 18 (2) : 177-181.

James, D.Y., E.C. Janice and M.R. Keith, (1982). Effects of chronic exposures to pesticides in animal systems (Eds. Jamice, E.C. and James, D.Y.) Raven Press, New York.

James, J.H., Ziparo, V., Jeppsson B. and Fischer, J.E.(1979). Hyperanemia, plasma amino acid imbalance and blood brain amino acid transport : A unified theory of portal systemic encephalopathy. *Lancet.* 2 : 772-775.

Juarez, P., Sanz, L. and Calvete, J.J. (2004). Snake venomics : characterization of protein families in *Sistrurus barbouri* venom by cysteine mapping, N-terminal sequencing, and tandem mass spectrometry analysis. *Proteomics.* **4** : 327-338.

Juarez, P., Wagstaff, S.C., Oliver, J., Sanz, L., Harrison, R.A. and Calvete, J.J. (2006). Molecular cloning of disintegrin-like transcript BA-5A from a Bitis arietans venom gland cDNA library: a putative intermediate in the evolution of the long-chain disintegrin bitistatin. *J. Mol. Evol.* **63**: 142-152.

Junaid Mahmood Alam. and Rashida Qasim. (1993). Changes in serum components induced by venoms of marine animals. *Pakistan. J. of Pharm. Sci.* **6(1)**: 81-87

Jurss, K. (1980). The effect of changes in external salinity on the free amino acids and two aminotransferases of white muscle from fasted *Salmo gairdneri R. Comparative Biochemistry and Physiology.* **65A** : 501-504.

Kabeer, A.S.I. Begum, Md.R., Sivaiah, S. and Ramana Rao, K.V. (1978). Effect of malathion on free amino acids, total proteins, glycogen and some enzymes of *L. marginalis. Proceedings of Indian Academy of Sciences.* **87**: 377-380.

Kamiguti, A.S., Theakston, R.D., Sherman, N. and Fox, J.W. (2000). Mass spectrophotometric evidence for p-iii/p-iv metalloproteinases in the venom of the boomslang *(dispholidus typus). Toxicon.* **38** : 1613–1620.

Karmarker, S.S., Barnett, R.J., Nicholas, M.M. and Seligman, A.M. (1959). Synthesis of P-nitro phenyl substituted tetrazolium salts containing iodine and other groups. *J. Am. Chem. So.* **81** : 3771-3778.

Karuzine, I.I. and Archakov, A. I. (1994). The oxidative inactivation of cytochrome P_{450} in monoxygenase reactions. *Free. Rad. Boil. Med.* **16**:73-97.

Kehrer, J.P., Jones, Lemasters, D. B., Farber, J.J. and Jarschke, K. (1990). Mechanisms of hypoxic cell injury. *Toxicol. Appl. Pharmacol.* **106** : 165-178.

Khessiba, A., Romeo, M. and Aissa, P. (2005). Effects of some environmental parameters on catalase activity measured in the mussel (*Mytilus galloprovincialis*) exposed to lindane. *Environ. Pollut.* **133**: 275-81.

King, E.J. and Wooton, I.D.P. (1959). In:Microanalysis in Medical Biochemistry, Churchill, London, P.42.

Kiran, K.M., More, S.S. and Gadag, J.R. (2004). Biochemical and clinicopathological changes induced by *Bungarus caeruleus* venom on a rat model. *Journal of clinical physiology and pharmacology.* **15 (3-4)** : 277-287.

Klassan, C.D. (1991). Heavy metal and heavy antagonists In : **Pharmacological basis of terapentics.** (L.S.Goodman, and Gilman. Eds.). London, Bailliere, Tindall. pp.1615.

Knox, W.E. and Greengard, O. (1965). The regulation of some enzymes of nitrogen metabolism on introduction to enzyme physiology. In: **Advances in enzyme regulation**. (G. Weber, and Bergman Eds.). Academic Press, New York. pp.247.

Kovacevic, Z. and Mc Givan, J.D. (1983). Mitochondrial metabolism of glutamine and glutamate and its physiological significance. *Physiol. Rev.* **63 (2)** : 547-605.

Kramer. and Tisdall, F.F. (1921).Estimation of serum potassium levels. *J. bio.chem.* **46:** 339.

Kularatne, S. A. M. and Ratnatunge, N. (2001). Autopsy study of common krait (*B. caeruleus*), *4th world congress of herpetology.* 59.

Kvamme, E. (1983). Deaminases and amides. In : **A hand book of neurochemistry** (Edited by Lajtha, A.). Plenum press, New York, **4** : pp. 405-422.

Latner, A.L. (1975). In : **Clinical Biochemistry**.(Cantaroiw and Trumper Eds.). VII Edn. W.B. Saunders Co., Philadelphia / London / Toronto, pp. 566.

Lee, Y.L. and Lardy, A.A. (1965). Influence of thyroid hormones on L-glycerophosphate dehydrogenases in various organs of the rat. *Journal of Biological Chemistry.* **240** : 1427-1430.

Lehninger., A.L. (1995). In : **Biochemistry** (Edited by Lybert stryer.), 4th edition. W.H. Free man and company, New york, pp. 41.

Leisuk, S.S., Czechowska, G., Zimmer, M., Slomka, S., Madro, M., Celinski, A. and Wielosz, M. (2003). Catalase, superoxide dismutase and glutathione peroxidase activities in various rat tissues after carbon tetrachloride intoxication. *J. Hepatobiol. Pancreat. Surg.* **10**: 309-315.

Li, Z.Y., Yu, T.F. and Lian, E.C.Y. (1994) Purification and characterization of L-amino acid oxidase from King Cobra (*Ophiophagus Hannah*) venom and its effects on human platelet aggregation. *Toxicon.* **32**, 1349-1358.

Litchfield, J.T. and Wilcoxon, F. (1949). A simplified method of evaluating dose-effect experiments. *J Pharmacol. Exp. Ther.* **96** : 99-1 13.

Liu, W., Li, M., Huang, F., Zhu, J., Dong, W. and Yang, J. (2006). Effects of cadmium stress on xanthine oxidase and antioxidant enzyme activities in Boleophthalmus pectinirostris liver. *Ying Yong Sheng tai Xue Bao.* **17(7)** : 1310 – 4.

Loeb, W.F. (1982). Clinical biochemistry of liver diseases. *Modified Veterinary Practice.* **63** : 625-631.

Lowenstein, J.M. (1972). The ammonia production in muscle and other tissues. The purine nucleotide cycle. *Physiol. Rev.* **52** : 382-414.

Lowry, O., Rosebrough, Farr, N.J. and Randall, R.J. (1951). Protein measurement with the folin phenol reagent. *Journal of Biological Chemistry.* **193** : 265-270.

Machlin, L.J. and Bendich, A. (1987). Free radical tissue damage Protective role of antioxidant nutrients. *FASEB.J.* **1**: 441-445.

Macht, D.I. (1938). Therapeutic experiences with cobra venom, *Ann Int Med.* **11** : 1824.

Mackessy, S.P., Baxter, L.M. (2006). Bioweapons synthesis and storage: the venom gland of front-fanged snakes. *Zoolog Anzeiger.* **245**: 147-159.

Manna, S., Bhattacharyya, D., Mandal, T.K. and Das, S. (2005). Repeated dose toxicity of deltamethrin in rats. *Indian Journal of Pharmacology.* **37(3)** : 160-164.

Marcinkiewicz, C., Weinreb, P.H., Calvete, J.J., Kisiel, D.G., Mousa, S.A., Tuszynski, G.P. and Lobb, R.R. (2003). Obtustatin: a potent selective inhibitor of alpha1beta1 integrin *in vitro* and angiogenesis *in vivo*. *Cancer Res.* **63** : 2020-2023.

Markland, F.S. (1983). Rattle snake venom enzymes that interact with components of the hemostatic system. *J.Toxicol.Toxin. Rec.***2**: 119.

Markland, F.S. Jr. (1997). Snake venoms. Drugs *54 Suppl.* **3**: 1-10.

Marsh, N., Gattullo, D., Pagliaro, P. and Losano, G., (1997) : The Gaboon viper, and Bitis gabonica : hemorrhagic, metabolic, cardiovascular and clinical effects of the venom. *Lif. Sci.* **61(8)** : 763-769.

Marsh, N.A. (1994). Snake venoms affecting the haemostatic mechanism – a consideration of their mechanisms, practical applications and biological significance. *Blood Coagul Fibrinolysis.* **5** : 399–410.

Match, D.I. (1936). Experimental and clinical study of cobra venom as analgesic, Proc Nat Acad Sci (USA). **22** : 61.

McCord, J.M. (1993). Human disease, free radical and the oxidant/ antioxidant balance in skeletal muscle after fatigue exercise. *J. Appl. Physiol.* **72** : 1111-1117.

McLane, M.A., Marcinkiewicz, C., Vijay-Kumar, S., Wierzbicka-Patynowski, I. and Niewiarowski, S. (1998). Viper venom disintegrins and related molecules. *Proc Soc Exp Biol Med.* **219** : 109-119.

Mebs, D. (2002). Venomous and poisonous Animals : In : **A Hand-book for Biologists, Toxicologists and Toxicologists, Physicians and Pharmacists** (Edited by Stuttgart.). CRC Press, Medpharm Scientific Publishers, pp.238-256.

Mehta, P.K., Hale, T.L. and Christen, P. (1993). Aminotransferases : demonstration of homology and division into evolutionary subgroups. *European Journal of Biochemistry.* **214** : 549-561.

Meier, J. and Stocker, K. (1991). Effects of snake venoms on hemostasis. *Crit Rev Toxicol.* **21** : 171-182.

Meier, J. and Theakston, R.D.Q. (1986). Approximate LD_{50} determinations of snake venoms using eight to ten experimental animals. *Toxicon.* **24 (4)** : 595-40.

Menez, A. (2002). In : **Perspectives in Molecular Toxicology** (Edited by Chichester), John Wiley and Sons, Ltd. United Kingdom, pp.123.

Menez, A., Stocklin, R. and Mebs, D. (2006). Venomics: The venomous systems genome project. *Toxicon.* **47**: 255-259.

Middleton, E.Jr. and Philips, G.B. (1964). Distribution and properties of anaphylactic induced slow reacting substance and histamine in guinea pigs. *J.Immunol.* **93** : 220-228.

Mohamed, A.H., Fouad, S., Abbas, F., Abdel Aal, A., Abdel baset, A., Amr Hassan, Abbas, N. and Zahran, F. (1981). Metabolic studies of the egyptian and allied african snake venoms. *Toxicon.* **18** : 381-383.

Molinengo, L. (1979). The curve doses *vs.* survival time in the evaluation of acute toxicity. *J. Pharm Pharmac.* **31** : 343.

Momeno, E., Alape, A., Sanchez, M. and Gutierez, J, M. (1988). A new method for the detection of phospholipase A_2 variants, identification of isoenzymes in the venoms of newborn and adult *Bothrops asper* (terciopelo) snakes. *Toxicon* . **26** : 363.

Mommsen, T.P. and Walsh, P.J. (1992). Biochemical and environmental perspectives on nitrogen metabolism in fishes. *Experientia.* **48** : 583-593.

Moore, S. and Stein, W.H. (1954). Modified ninhydrin reagent for the photometric determination of amino acids and related compounds. *Journal of Biological Chemistry.* **221** : 907-913.

Moorthy, K.S., Kasi Reddy, B., Swami, S.K. and Chetty, C.S. (1984). Changes in respiration and ionic content in tissues of freshwater mussel exposed to methylparathion toxicity. *Toxicology Letters.* **21** : 287-291.

Morita, T. (2004). Use of snake venom inhibitors in studies of the function and tertiary structure of coagulation factors. *Int. J. Hematol.* **79** : 123–9.

Moser, V.C. (1990). Approaches for assessing the validity of functional observational battery. *Neurotoxical, Teratol.* **12** : 483- 488.

Moustafa, F.A., Ahmed, Y.Y. and El-Asmar, M.F. (1974). Effect of scorpion (*Buthus minax L. Koch).* venom on succinic dehydrogenase and cholinesterase activity of mouse tissue. *Toxicon.* **12** : 237-240.

Mukhopadhyay, P.K., Mukherji, A.P. and Dehadrai, P.V.(1982). Certain biochemical responses in the air-breathing catfish *Clarias batrachus* exposed to sublethal carbofuran. *Toxicology.* **23** : 337-345.

Murray, Robert K., Daryl, K., Granner, Peter, A., Mayes. and Victor, W., Rodwell, (2007). In: **Harper's Illustrated Biochemistry.** International 26th Edition. McGraw-Hill Company Inc, New York, pp. 47.

Nachlas, M.M., Margulius, S.D. and Seligman, A.M. (1960). A colorimetric method for estimation of succinic dehydrogenase activity. *Journal of Biological Chemistry.* **235** : 499-504.

Natelson, S. (1971). Total cholesterol procedure (Libermann Burchard reagent) or free fatty acids in serum. In : **Techniques of clinical chemistry,** (Edited by Charles. C.). Thomas Publishers, Spring field Illinois, U.S.A, pp. 263-268.

Nelson, D.L. and Cox, M.M. (2005). In : **Lehninger Principles of Biochemistry,** Fourth Edition, W.H. Freeman and Company, New York, pp.178.

Nelson, M., Lopera-Barrero, Jayme, Povh, A., Ricardo, P., Ribeiro, Patricia, C., Gomes, Carolina, B., Jacometo. and Tais da Silva Lopes. (2008). *Cien. Inv. Agr.* **35 (1)** : 65-74.

Oja, S.S., Von Bonsdorff, H.A.R. and Lindoors, O.F.C. (1966). Ammonia content of developing rat brain. *Nature.* **212** : 937-938.

Omotuyi, I., Oluyemi, K.A., Omofoma, C. O., Josaiah, S. J., Adsanya, O. A. and Saalu, L.C. (2006). Cyfluthrin induced hepatotoxicity in rats. *African Journal of Biotechnology.* **5 (20)** : 1909-1912.

Oruc, E.O. and Uner, N. (1999). Effect of 2,4-Diaminon some parameters of protein and carbohydrate metabolism in the serum, muscle and liver of *Cyprinus carpio. Environmental Pollution.* **105 (2)** : 267-272.

Pallabi, D.E. (2000). Pharmacological and toxicological studies on the Indian king cobra (*Ophiophagus Hannah*) venom. Ph.D. Thesis, Calcutta University, Kolkata, India.

Pasquier, C., Pierce, B., Willson, R.T., Hyyaishi, O., Niki, E., Kondo, M. and Yoshikova, T. (1989). Xanthine oxidase mediated free radical mediated injury . *Med. Biochem. Chem. Aspects. Free. Radic.* 425-432.

Peichoto, M.E., Teibler, P., Ruiz, R., Leiva, L. and Acosta, O. (2006). systemic pathological alterations caused by *Philodryas patagoniensis* collubrid snake venom in rats. *Toxicon.* **48** : 520–528.

Pellegrino, C. and Franzini, C. (1963). An electron microscopic study of denervation atrophy in red and white skeletal muscle fibers. *Cell Biol.* **17** : 327.

Picolo, G., Giorgi, R., Bernardi, M.M. and Curry, Y. (1998). The antinociceptive effect of *Crotalus Durissus Terrificus* snake venom is mainly due to a supraspinally integrated response. *Toxicon.* **36**: 223.

Pradeep Kumar, K.M. and Basheer, M.P. (2011). Snakebite: Biochemical changes in blood after envenomation by viper and cobra. *J.Med. Allied Sci.* **1(1)**: 36- 41.

Rabie, F., El-Asmar, M.F. and Ibrahim, S.A. (1972). Inhibition of catalase in human erythrocytes by scorpion venom, *Toxicon.* **10** : 87-88.

Radhakrishnaiah, K., Suresh, A., Urmila Devi, B. and Sivaramakrishnaiah, B. (1991). Effect of mercury on the lipid metabolic profiles in the organs of *Cyprinus carpio* (Linnaeus). *J. Mendel.* **8** : 123-125.

Rafat Yasmees, M. (1986). Physiological responses of freshwater fish, *Anabas scandens* to the toxicity of endosulfan. Ph.D. Thesis, Osmania University, Hyderabad, India.

206

Rahmy, T.R., Ramadan, R.A., Farid, T.M. and El-Asmar, M.F. (1995). Renal lesions induced by cobra envenomation. *J. Egypt. Ger. Soc. Zool.* **17(C)**: 251-271.

Ramanadikshithulu, Narayana Reddy, A.V. and Swamy, K.S. (1976). Effect of selected metal ions on glutamate dehydrogenase activity in cell free extract of goat liver. *Indian Journal of Experimental Biology.* **14** : 621-623.

Rathore, N., Kale, M., John, S. and Bhatnagar, D. (2000). Lipid peroxidation and anti oxidant enzymes in isoproterenol induced oxidative stress in rat erythrocytes. *Indian. Physiol. Pharmacol.* **44 (2)** : 161 – 166.

Ray, G. and Hussain, S.A. (2002). Oxidants, antioxidants and carcinogenesis. *Ind. J. Exper. Biol.* **40** : 1213-1232.

Reed, L.J. and Muench, M. (1958) A simple method of estimating fifty percent end product. *Am J. Iyg.* **27** : 495-497.

Rahman A., Nagi, A.H. and Hayee, A. (2006). Effects of prolonged poisoning by cobra venom on blood coagulation, platelets and fibrinolysis. *Biomedica.* **22** :11.

Reitman, S. and Frankel, S. (1957). A colorimetric method for the determination of glutamine oxaloacetic acid, glutamic pyruvate transaminases. *American Journal of Clinical Pathology.* **28** : 56-63.

Roncaglioni, M.C., Gde Gaetano, and M.B. Donati. (1982). Animals in toxicological research. Instituo di Recherche. *Farmocologiche.* **14***:* 77-89.

Rosenberg, P. (1979). Pharmacology of Phospholipase A_2 from snake venoms. In : **Snake venoms (Handbook of Experimental Pharmacology)**, (Edited by Lee, C. Y.). Springer – Verleg, Berlin, pp.403.

Rus, D.A., Sastre, J., Vina, J. and Pallardo, F.V. (2007). Induction of mitochondrial xanthine oxidase activity during apoptosis in the rat mammary gland. *Front. Biosci.* **1** : 1184.

Ryrfeldt, A., Bennenberg, G. and Moldeus, P. (1992). Free radicals and lung disease. In : **Free radicals in medicine** (Cheeseman, K.H, Slater T.F. Eds.). Church Hill Livingstone, London. pp. 588-603.

Sakaguchi, Y., Yuge, K., Yoshino., Yamashita, M. F. F. and Hashimoto, T. (1981). In: **Hyperammonemia in the neonate with hypoxia in urea cycle diseases.** (Lowenstein, A., Mori, A. and Marescan, B. Eds.).**153**: 147.

Samejima, Y., Aoki, Y., and Mebs, D. (1988). Structural studies on myotoxin from *Crotalus adamanteus* venom. In: Progress in venom and toxin research, Gopalakrishnaone, G., and Tan, C.K. Eds., National University of Singapore, Singapores, pp. 186.

Samuel, (1977). In : **Notes on Clinical Lab Techniques.** M.K.G.Iyyer and son publishers, Chennai.

Sant, S.M. and Purandare, N.M. (1972). Autopsy study of cases of snake bite with special reference to renal lesions. *J. Postgrad. Med.* **18** : 181-88.

Schmike, R.T. (1974). In: **Neuro Sciences** (Edited by Schmitt and F.G. Worder.). Cambridge, Massachusetts II, pp.813-825.

Shakoori, A.K., Zaheer, S.A. and Ahmed, M.S. (1976). Effect of malathion, dieldrin and endrin on blood serum protein and free amino acid pool of *Channa punctatus* (bloch). *Pakistan Journal of Zoology.* **8** : 125-134.

Shier, W.T., Eaker, Wadstrom, D., Pergamon, T., Eds. (1980). Activation of self-destruction as a mechanism of action for cytolytic toxins. In : *Natural Toxins.* Oxford, pp.193.

Shiomi, K., Yamaoto, S., Yamanaka, H., Kikuchi.T. and Konno, K. (1990). Liver damage by the crown-of- thorns starfish (*Acanthaster planci)* lethal factor. *Toxicon.* **28** : 469-475.

Singaraju, R., Subramanian, M.A. and Varadaraju, G. (1991). Sublethal effect of malathion on the protein metabolism in the fresh water field crab *paratelphusa hydrodromous. Ecotoxicol. Environ. Monit.* **1(1)** : 41-44.

Sitprija, V. (2006). Snake bite nephropathy. N*ephrology.* **11** : 442–448.

Sitprija, V. and Boonpucknaving, V. (1977). The kidney in tropical snakebite. *Clin. Nephrol.* **8** : 377-383.

Sitprija, V., Suvanpha, R., Pochanugool, C., Chusil, S. and Tungsanga, K., (1982). Acute interstitial nephritis in snakebite. *American Journal Tropical Medicine and Hygiene.* **31** : 408-410.

Snedecor, G. W. and Cochran, W. G. (1968). In: **Statistical Methods**. 6th edition. Oxford and IBH publishing company, Calcutta, Bombay and New Delhi, pp.168-181.

Soe, S., Win, Htwe, M.M., Lwin, T.T., Thet, M. and Kyaw, W.W. (1993). Renal histopathology following Russels viper (*Vipera russelli*) bite. *South East Asian journal of Tropical medicine and public health.* **24**: 193-197.

Souza, D.H., Eugenion, L.M., Fletcher, J.E., Jiang, M.S., Garratt, R.C., Oliva, G. and Selistre-Araujo, H.S. (1999). Isolation and structural characterization of a cytotoxic L-amino acid oxidase from *Agkistrodon contortix laticimetus* snake venom : preliminary crystallographic data. *Arch. Biochem. Biophy.* **368 (2)** : 285-90.

Sreedevi, P., Suresh, A., SivaRamakrishna, B., Prabhavathi. And Radhakrishnaiah, (1992). Bioaccumulation of nickel in the organs of fresh water fish *Cyprinus carpio* and fresh waters mussel *Lamellidens marginalis. Chemosphere.* **24** : 29-36.

Srikanthan, T.W. and Krishna Murthy, C.(1955). Tetrazolium test for dehydrogenases. *Sci. Ind. Res.* **14** : 206.

Stagni, N. and Debernard, B. (1968). Lysosomal enzyme activity in rat and beef skeletal muscle. *Biochem. Biophys. Acta.* **170 (1)** : 129-139.

Stocker, K. (1990). Composition of snake venoms. In : **Medical use of snake venom proteins**, (Edited by Stocker, K.F.) CRC Press, Boca Raton, pp.33-56.

Sund, H., Dieter, H. R., Koberstein, R. and Rasched, I. (1977). G.D.H : Chemical modification and ligand binding. *Journal of Molecular Catalysis.* **2** : 1-23.

Tavill, A.S. and W.G.S. Cooksly. (1983). In: **Biochemical aspects of liver disease.** (Elkeles, R.S. and Tavil, A.S. Eds.). Black Well Scientific Publications, Boston, pp.144.

Temel, I., Bay, E., Cigli, O. and Akyol, O. (2002). Erythrocyte catalase activities in alcohol consumption, medications and some diseases. *Inonu. Univer. Derg.* **9(1)** : 373-382.

Tilbury, R.C., Madkour, M.M., Saltissi, D. and Suleiman, M. (1987). Acute renal failure following the bite of burton's carpet viper *Echis coloratus* gunther in Saudi Arabia: Case report review. *Saudi Med J.* **8** : 87-95.

Trevan, J.W. (1927). The error of determination of toxicity. *Proc R. Soc.* **101 B** : 485.

Trinder, P. (1915). Estimation of sodium.*Analyst.* **76** : 596.

Tu, A. R. (1988). Overview of snake venom chemistry. *Adv. Exp. Med. Boil.* **391** : 37-62.

Tu, A., Ed. (1977). Nonneurotoxic basic proteins (cardiotoxins, cytotoxins, and others) In : **Venoms : Chemistry and Molecular Biology**. John Willey, New York, pp.301.

Turner, J.C. and V. Shanks. (1980). Absorption of some organochlorine compounds by rat small intestine *in vivo. Bulletin of Environmental Contamination and Toxicology.* **24** : 625.

210

Ueda, M., Change, C.C. and Ohno, M. (1988). Purification and characterization of L-amino acid oxidase from the venom of *Trimeresurus mucrosquamatus* (Taiwan habu snake) venom. *Toxicon.* **26** : 693.

Usha Rani, (2010). Synergetic impact of chorpyrifos and stocking density stress on *catla catla* (Hamilton) and *labeo rohita* (Hamilton) with reference to protein metabolism and histopathology. Ph.D Thesis, Sri Venkateswara University, Tirupati. India

Vani, M. (1991). Involvement of liver in detoxification mechanism in albino rat under sublethal doses of chlordane an OC compound. Ph.D. Thesis Sri Venkateswara University,Tirupati. India

Vaziri, N.D., Lin, C.Y., Farmand, F. and Ram, K.S. (2003). Superoxide dismutase, catalase, glutathione peroxidase and NADPH oxidase in lead induced hypertension in Kidney. *Intl.* **36** : 186-194.

Venkataiah, A. (1995). Effect of skeletal exercise training on age related antioxidant defense mechanism in rat skeletal muscles. PhD Thesis, Sri Venkateswara University, Tirupati, India.

Venkataswamy, K. (1991). Neurochemical studies during the development of behavioral tolerance to organophosphate toxicity in albino rats. Ph.D. Thesis, Sri Venkateswara University, Tirupati, India.

Venkateswarulu, D., Surendra Reddy, W.V. and Sasira Babu, W. (1978). Effect of the scorpion *Heterometrus fulvipes (C.Koch)* venom on some enzyme systems in rat (albino) tissues. *Exp.* **34** : 233-234.

Vorbach, C., Harrison, R. and Capecchi, M.R. (2003). Xanthine oxido reductase is central to the evolution and function of the innate immune system. *Trends in Immunol.* **24 (9)** : 512-517.

Waarde, A.V. and Kesbeke, F. (1982). Nitrogen metabolism in gold – fish *Carassicus auratus* (L). Influence of added substrates and enzyme inhibitors on ammonia production of isolated hepatocytes. *Comparative Biochemistry Physiology.* **70 (B)** : 499-507.

Walton, J.M. and C.B. Cowey. (1982). Aspects of intermediary metabolism in Salmonid fish. *Comparative Biochemistry Physiology.* **73 (B)** : 59-79.

Wootton, I.D.P. (1964). In: **Micro analysis in medical biochemistry**. churchil, livingstone, pp.150.

Wootton, I.D.P. (1974). In: **Micro analysis in Medical Biochemistry**, Churchill, living stone, pp.154.

Zahra, F.H., Mohammad, F., Mojtaba, T.K. (2005). Effects of *Echis carinatus* venom on the haemodynamy and contractility of vascular and visceral smooth muscle of rats. *Toxicology Mechanisms and Methods.* **15 (1)** : 53-57(5).

Zbinden, G. and Flurry. (1981). Significance of the LD_{50} test for the toxicological evaluation of chemical substances. *Arch. Tox.* 47-77.

Zhao, X., Yeh, J.Z. and Narahashi, T. (2001). Post-stroke dementia. Nootropic drug modulation of neuronal nicotinic acetylcholine receptors. *Ann N Y Acad Sci* . **939** : 179-186.

www.ingramcontent.com/pod-product-compliance
Lightning Source LLC
Chambersburg PA
CBHW070230190526
45169CB00001B/143